Dynamic Flavor: Capturing Aroma Using Real-Time Mass Spectrometry

ACS SYMPOSIUM SERIES **1402**

Dynamic Flavor: Capturing Aroma Using Real-Time Mass Spectrometry

Jonathan D. Beauchamp, Editor

Department of Sensory Analytics and Technologies
Fraunhofer Institute for Process Engineering and Packaging IVV
Freising, Germany

Sponsored by the
ACS Division of Agricultural and Food Chemistry, Inc.

American Chemical Society, Washington, DC

Library of Congress Cataloging-in-Publication Data

Names: Beauchamp, Jonathan, editor.
Title: Dynamic flavor : capturing aroma using real-time mass spectrometry / Jonathan D. Beauchamp, editor, Department of Sensory Analytics and Technologies, Fraunhofer Institute for Process Engineering and Packaging IVV, Freising, Germany.
Description: Washington, DC : American Chemical Society, [2021] | Series: ACS symposium series ; 1402 | "Sponsored by the ACS Division of Agricultural and Food Chemistry, Inc." | Includes bibliographical references and index.
Identifiers: LCCN 2021051779 (print) | LCCN 2021051780 (ebook) | ISBN 9780841297944 (hardcover) | ISBN 9780841297937 (ebook other)
Subjects: LCSH: Food--Sensory evaluation. | Flavor. | Odors. | Taste--Molecular aspects. | Smell--Molecular aspects. | Mass spectrometry.
Classification: LCC TX546 .D96 2021 (print) | LCC TX546 (ebook) | DDC 664/.07--dc23/eng/20211203
LC record available at https://lccn.loc.gov/2021051779
LC ebook record available at https://lccn.loc.gov/2021051780

The paper used in this publication meets the minimum requirements of American National Standard for Information Sciences—Permanence of Paper for Printed Library Materials, ANSI Z39.48n1984.

Copyright © 2021 American Chemical Society

All Rights Reserved. Reprographic copying beyond that permitted by Sections 107 or 108 of the U.S. Copyright Act is allowed for internal use only, provided that a per-chapter fee of $40.25 plus $0.75 per page is paid to the Copyright Clearance Center, Inc., 222 Rosewood Drive, Danvers, MA 01923, USA. Republication or reproduction for sale of pages in this book is permitted only under license from ACS. Direct these and other permission requests to ACS Copyright Office, Publications Division, 1155 16th Street, N.W., Washington, DC 20036.

The citation of trade names and/or names of manufacturers in this publication is not to be construed as an endorsement or as approval by ACS of the commercial products or services referenced herein; nor should the mere reference herein to any drawing, specification, chemical process, or other data be regarded as a license or as a conveyance of any right or permission to the holder, reader, or any other person or corporation, to manufacture, reproduce, use, or sell any patented invention or copyrighted work that may in any way be related thereto. Registered names, trademarks, etc., used in this publication, even without specific indication thereof, are not to be considered unprotected by law.

Foreword

The ACS Symposium Series is an established program that publishes high-quality volumes of thematic manuscripts. For over 40 years, the ACS Symposium Series has been delivering essential research from world leading scientists, including 36 Chemistry Nobel Laureates, to audiences spanning disciplines and applications.

Books are developed from successful symposia sponsored by the ACS or other organizations. Topics span the entirety of chemistry, including applications, basic research, and interdisciplinary reviews.

Before agreeing to publish a book, prospective editors submit a proposal, including a table of contents. The proposal is reviewed for originality, coverage, and interest to the audience. Some manuscripts may be excluded to better focus the book; others may be added to aid comprehensiveness. All chapters are peer reviewed prior to final acceptance or rejection.

As a rule, only original research papers and original review papers are included in the volumes. Verbatim reproductions of previous published papers are not accepted.

ACS Books

Contents

Preface ix

1. **Non-destructive and High-Throughput—APCI-MS, PTR-MS and SIFT-MS as Methods of Choice for Exploring Flavor Release** 1
   Andrew J. Taylor, Jonathan D. Beauchamp, and Vaughan S. Langford

2. **Flavor Applications of Direct APCI-MS** 17
   Andrew J. Taylor

3. **Pushing the Boundaries of Dynamic Flavor Analysis with PTR-MS** 33
   Jonathan D. Beauchamp

4. **SIFTing Through Flavor—Exploring Real-Time, Real-Life Processes Using SIFT-MS** 51
   Diandree Padayachee and Vaughan S. Langford

5. **Understanding the Molecular Basis of Aroma Persistence Using Real-Time Mass Spectrometry** 67
   Carolina Muñoz-González, María Ángeles Pozo-Bayón, and Francis Canon

6. **Using APCI-MS to Study the Dynamics of Odor Binding under Simulated Peri-Receptor Conditions** 77
   Andrew J. Taylor and Masayuki Yabuki

7. **APCI-MS/MS—An Enhanced Tool for the Real-Time Evaluation of Volatile Isobaric Compounds** 87
   Ni Yang, Clive Ford, and Ian Fisk

8. **From Mold Worms to Fake Honey: Using SIFT-MS to Improve Food Quality** 99
   Sheryl Barringer

9. **Identifying Potential Volatile Spoilage Indicators in Shredded Carrot Using SIFT-MS** 107
   Lotta Kuuliala, Nikita Jain, Bernard De Baets, and Frank Devlieghere

10. **Real-Time Monitoring of Flavoring Starter Cultures for Different Food Matrices Using PTR-MS** 123
    Vittorio Capozzi, Mariagiovanna Fragasso, Iuliia Khomenko, Patrick Silcock, and Franco Biasioli

11. **Monitoring of Acrolein, Acetaldehyde and 1,3-Butadiene in Fumes Emitted during Deep-Frying of Potato Pieces in Rapeseed Oil Using PTR-MS** 139
    Wiktoria Wichrowska and Tomasz Majchrzak

Editor's Biography **151**

Indexes

Author Index.. **155**

Subject Index **157**

Preface

Exploring the dynamic processes involved in flavor release and perception calls for the use of suitably fast analytical tools. Flavor research has historically centered around the extraction and capture of discrete flavor scenarios, followed by the analysis and subsequent orchestration of these individual snapshots to compile an overall picture of flavor composition, or in the case of dynamic processes, to provide an overview of (potential) changes over time. While delivering comprehensive data that has incontrovertibly furthered our knowledge of flavor, such approaches are laborious and time-consuming, and are certainly not conducive to examining fast processes. The development of real-time mass spectrometry techniques at the end of the twentieth century, notably atmospheric pressure chemical ionization-mass spectrometry (APCI-MS), proton transfer reaction-mass spectrometry (PTR-MS), and selected ion flow tube-mass spectrometry (SIFT-MS), opened fresh possibilities for exploring dynamic flavor processes and represented a new era in food/flavor research. With their fast sub-second analytical responses, broad dynamic range and sensitive detection of individual volatile compounds, these real-time mass spectrometry techniques – often referred to as direct injection mass spectrometry (DIMS), amongst other terms – were ideally catered to investigating rapid flavor processes, from characterizing breath-by-breath aroma release *in vivo*, to monitoring the kinetics of food processing, spoilage or maturation, or analyzing food headspace non-destructively *in vitro* with high-throughput. The implementation of these tools in the quarter century since their inception has amassed a wealth of data on diverse aspects of flavor release that has furthered our understanding of the dynamic processes involved.

The ensuing pages of this volume offer an overview of notable achievements and novel applications of these DIMS tools in the area of food/flavor science, presented in eleven chapters comprising two general chapters and three chapters each on APCI-MS, PTR-MS and SIFT-MS. The book commences with an introductory overview of the three technologies, from their early developments and uses, to their operating principles and individual nuances. The next three chapters each focus on a particular technique, highlighting key applications that showcase their strengths. Subsequent chapters are a mix of reviews and experimental research relating to the use of these DIMS tools for a range of topics. These include exploring flavor perception, aroma persistence and odor binding via *in vivo* or *in vitro* analyses, examining spoilage, fermentation or food authenticity via dynamic headspace or high-throughput sampling in combination with powerful chemometrics data processing techniques, developing new technology-based strategies to aid compound identification, and characterizing flavor generation online during food processing. Collectively, these works serve the purpose of providing a broad overview of the capabilities and novel areas of implementation of these technologies and offer a glimpse of prospective applications on the horizon.

As a secondary aspect – and an intended goal of this undertaking – this book brings together three tools that have experienced a contentious history. Irrefutably, innovation is a driver of progress, as is clearly evident from the new insights on dynamic flavor processes gained by the emergence and implementation of real-time mass spectrometry. Equally, competition between these technologies has led to advancements and differentiating capabilities of each tool, offering tailored strengths for one particular application or another. Ultimately, however, any analytical instrument is only a means to an end, providing the necessary features to make it possible to examine particular processes. Across these pages, the overall capabilities of real-time mass spectrometry shine through, irrespective of the specific technology under scrutiny. Each instrument exhibits unique features that offer benefits over the other two, but it is their common strengths, and the shared ambitions of the chapter authors to undertake state-of-the-art research, that should resonate with the reader. Innovation drives progress, but discourse nurtures understanding.

As should be immediately apparent from the book series in which this work is published, the roots of this volume lie in an American Chemical Society (ACS) symposium organized within the Division of Agricultural & Food Chemistry (AGFD). The symposium in question, entitled *Food-Flavor Dynamics Assessments via Real-Time Mass Spectrometry*, was coordinated by myself and Yu Wang and was held at the ACS Spring 2021 National Meeting, which took place April 5-16, 2021 in a virtual format on account of global travel restrictions and related pandemic-induced sensibilities. Split into two sessions, *From Development to Nosespace* and *Headspace and Beyond*, the symposium comprised sixteen oral presentations by speakers participating from different corners of the world, from North America across Europe to New Zealand. Like this subsequent book, the intention of the symposium was to bring together food researchers actively using real-time mass spectrometry for novel applications towards a common goal of understanding dynamic processes in relation to flavor, and to thereby offer a platform for scientific discourse on this complex and captivating topic.

There are several people whom I would like to thank for playing key roles in the materialization of this book. Chronologically, the idea for the related ACS symposium came about through discussions with Brian Guthrie and Michael Morello, both of whom shared their enthusiasm for the topic and offered their encouragement to organize this session; without this impetus, the symposium might never have transpired. As with many endeavors, sharing the workload is often a most welcome relief; I am very grateful to my co-organizer, Yu Wang, for her help in putting the symposium together and for her skilled moderation of the sessions. The success of the symposium largely rests on the shoulders of the presenters, many of whom committed to contributing a chapter to this book; accordingly, I would like to thank, in order of appearance, Christian Lindinger, Diandree Padayachee, Carolina Muñoz, Jean-Luc Le Quéré, Iuliia Khomenko, Leonardo Menghi, Michele Pedrotti, Sheryl Barringer, Joseph Timkovsky, Patrick Silcock, Patrik Španěl, Susana Ratnawati and Tomasz Majchrzak, who were not discouraged by the conference scheduling that spanned seventeen time zones and resulted in mostly late-night or crack-of-dawn presentations, albeit from the comfort of their own homes!

Of course, the book would be empty were it not for the contributions of the authors, whom I acknowledge here without naming them individually (cf. the contents page), with two exceptions. First, special thanks is due to Andy Taylor, who not only gave two talks at the symposium and contributed two chapters in this work on APCI-MS, but also offered advice and gave assistance in putting this volume together. Further, he took the lead in writing the first chapter of this book, which introduces the three key real-time MS technologies – APCI-MS, PTR-MS and SIFT-MS – that acts as a basis for subsequent chapters. Second, I would like to thank Vaughan Langford for agreeing to additionally co-author this introductory chapter and for contributing his extensive knowledge on SIFT-MS, as well as his keen eye for detail; the resulting chapter is an excellent example of how proponents of rivalled technologies can collaborate to put together a (hopefully unbiased) review of the individual capabilities and common benefits offered by these tools.

Final thanks are directed at Youngmok Kim, current AGFD Chair, for his support and encouragement, as well as the staff at ACS Publications, notably Amanda Koenig, for initiating this book and for her work behind the scenes, Maryanne Rackl, for her tireless efforts (and composure) in processing all the chapters, and Alison Kreckmann, for her talents on the book cover design.

If you have picked up this volume then either you are interested in the dynamic processes of flavor development, or you are flavor analyst already using one of the real-time mass spectrometry techniques covered in these pages, or possibly both. In any case, I hope this book serves as an interesting and enlightening reference work for related studies that will further push the boundaries of our understanding of dynamic flavor processes.

October 2021

Jonathan D. Beauchamp
    Fraunhofer Institute for Process Engineering and Packaging IVV
    Department of Sensory Analytics & Technologies
    Giggenhauser Str. 35
    85354 Freising
    Germany

Chapter 1

Non-destructive and High-Throughput—APCI-MS, PTR-MS and SIFT-MS as Methods of Choice for Exploring Flavor Release

Andrew J. Taylor,[1] Jonathan D. Beauchamp,[2] and Vaughan S. Langford[3,*]

[1]Flavometrix Ltd., Loughborough, Leicestershire, United Kingdom
[2]Fraunhofer Institute for Process Engineering and Packaging IVV, Department of Sensory Analytics & Technologies, Giggenhauser Str. 35, 85354 Freising, Germany
[3]Syft Technologies Ltd., 68 Saint Asaph St., Christchurch 8011, New Zealand
*Email: vaughan.langford@syft.com

Exploring the dynamic nature of aroma release from foods demands analytical tools capable of fast and sensitive measurements. Real-time mass spectrometry represents a class of technology that offers rapid detection of volatile organic (aroma) compounds (VOCs) at trace concentrations. The emergence of these instruments almost three decades ago opened up the field of food science to closely examine the relationship between food aroma composition and flavor perception in a direct and continuous manner, offering a complementary means to the established conventional approach using gas chromatography-mass spectrometry (GC-MS). Three technologies in particular exhibited suitable capabilities to fulfil the analytical demands for real-time analysis of VOCs and thus dominated this new arena, where they remain the driving technologies today: atmospheric pressure chemical ionization-mass spectrometry (APCI-MS), proton transfer reaction-mass spectrometry (PTR-MS), and selected ion flow tube-mass spectrometry (SIFT-MS). These systems have afforded extensive studies on flavor release, from headspace analysis to *in vivo* aroma release during food consumption. This chapter introduces these technologies and provides an overview of their key common and differentiating features as a lead-in to the subsequent review and applications chapters of this book.

Rationale for Real-Time Mass Spectrometry to Study Aroma Release

For many years, flavor researchers have studied the links between the volatile compounds that are present in foods and the perception of food aroma in the consumer. Up until the mid-1990s, the focus was on identifying and quantifying the odorous volatile components in a wide range of

© 2021 American Chemical Society

foods by extracting the compounds, typically with an organic solvent, and analyzing the extract using gas chromatography-mass spectrometry (GC-MS). While this comprehensive approach was key to understanding flavor composition, attempts at re-creating the aroma by combining the volatile compounds in a model system (at the ratios measured in the real food) were not always successful. It became clear that the food matrix plays an important role in the way aroma compounds were released from the product before and during eating, and that factors such as fat content or viscosity could significantly affect the timing and the degree of aroma release. Since humans detect and perceive aromas after they have been released from the food, the focus of flavor studies turned to the analysis of food headspace, i.e., the volatiles released from the food into the air above the product. Analyses of the headspace aroma composition using GC-MS were carried out at equilibrium, for example, by placing a food sample in a sealed vessel, usually a capped glass vial, and allowing time for the aroma compounds to partition between the food and the air phases. A sample of air was then analyzed by GC-MS. While this gives useful information, in real life, aromas are rarely found at equilibrium. For example, the aroma profile during the coffee brewing process, or as food is chewed by consumers, changes considerably and so a technique to measure aroma composition under these dynamic conditions was needed.

Aroma release during eating is a fairly rapid process. Volatile compounds are released from the food matrix into the air in the oral cavity as the food undergoes mechanical breakdown during chewing and interacts with saliva. Breathing and swallowing then transfer these volatile aroma compounds to the olfactory receptors via the nasopharynx, in what is referred to as the retronasal pathway (*1*). The process occurs over periods between 30 s to 2 min, with about 12 breaths per minute for the average human. Monitoring aroma release using chromatographic techniques involves a lot of work to sample the aroma compounds on each breath and subsequently analyze individual samples (*2*). Direct methods of analysis capable of continuously monitoring aroma release were an obvious solution to this research challenge. As the new millennium approached, researchers in other fields were also interested in monitoring volatile compounds from breath, e.g., for potential health applications, and real-time mass spectrometry was a highly attractive technique. In this approach, gas samples are introduced to, and analyzed by, the system directly and immediately, with no chromatography. Examples from this era include the work of Benoit *et al.* (*3*), who explored real-time breath analysis as a potential diagnostic method for monitoring volatile compounds in exhaled breath using atmospheric pressure chemical ionization-MS (APCI-MS) and Soeting and Heidema (*4*), who measured aroma release from aqueous solutions during simulated eating. These, and other systems, were partially successful. In the case of breath analysis by APCI-MS (*3*), the dead volume in the interface linking exhaled breath to the mass spectrometer was large and therefore the response was too slow for time-resolved breath-by-breath analyses. By comparison, the membrane system that separated the atmospheric pressure of the breath inlet and the high vacuum present in the mass spectrometer (*4*) adsorbed some compounds, resulting in different transmission times to the detector. During the same period, parallel developments of other real-time mass spectrometry techniques to measure volatiles at low concentrations were underway, including selected ion flow tube-MS (SIFT-MS), with a focal application on health-related breath analysis (*5*), and proton transfer reaction-MS (PTR-MS) (*6*), which explored different potential fields of application, from atmospheric chemistry analyses, to medical breath analysis applications, and food aroma release. Unlike the direct ionization of target analytes at the corona discharge source in APCI-MS, both PTR-MS and SIFT-MS utilize separate front-end ion sources to generate reagent ions that are subsequently injected into a reaction cell, whereupon they initiate chemical ionization with volatile compounds in the sample gas that enter this region via a transfer line. The well-defined conditions in these reaction chambers – referred to as a flow tube in SIFT-MS and a flow-drift tube in PTR-MS (see later) – allowed for studies on reaction rate coefficients between reagent ions and neutral analytes, which could be utilized for pseudo-absolute quantitation of trace compounds based on kinetic theory.

In the early days of exploring the potential of these real-time mass spectrometry techniques for dynamic aroma release studies – sometimes referred to as direct injection mass spectrometry (DIMS) – it soon became apparent that various common factors needed to be taken into account. These factors (reviewed in detail elsewhere (7)) can be broadly categorized as:

- Interfacing humans with the instrument
- Optimizing and controlling ionization
- Deconvoluting the complex mass spectra to match compounds to ions
- Experimental design and data processing

In short, it was necessary to design safe and reliable systems that could sample exhaled air from humans as they ate food and ensure that the volatile compounds were rapidly transported to the ionization system without condensing in, or being adsorbed by, the materials of the transfer line, as well as to consider how ionization would be affected when multiple compounds were simultaneously introduced into the ion source or reaction cell. This was in great contrast to GC-MS, which separates compounds first and typically introduces single compounds into the ionization source from which clear spectral "fingerprints" can be obtained. Sensitivity of the detector was also an issue, as some aroma compounds are perceived by humans at around the parts per trillion by volume (ppt_V) level, which is close to the limit of detection of such instruments.

Following the early developments of real-time mass spectrometry, the three aforementioned techniques emerged as being especially suited to exploring the dynamic processes of flavor release. Although other techniques exist, this chapter focuses on the most notable and widely implemented technologies, namely APCI-MS, PTR-MS and SIFT-MS, which remain the most common approaches for flavor release studies. One factor to emphasize is that real-time mass spectrometry detects individual compounds according to the mass-to-charge (m/z) ratio of the ensuing product ions, whereby these ions are always singly charged, i.e., $|z|=1$. Accordingly, these real-time techniques are not designed to identify compounds and are best suited for targeted analysis; therefore, ideally the aroma composition should be ascertained beforehand by GC-MS to provide complementary, confirmatory data on compound identities.

Overview of Real-Time Mass Spectrometry Methods

A common feature of APCI-MS, PTR-MS and SIFT-MS is that target volatiles are ionized via chemical ionization (CI), which was first discovered by Munson and Field in the mid-1960s (8). Unlike electron ionization (EI), which involves ionization at 70 eV, CI reactions proceed with energies as little as 1-2 eV (9). Accordingly, this form of ionization is referred to as "soft" ionization. The advantage of the low energy reactions afforded by CI is its less destructive nature, whereby the degree of fragmentation of the target analyte upon ionization is minimal and often the intact molecular (protonated) ion represents the dominant feature in the mass spectrum. All three techniques commonly use protonated water (H_3O^+) – technically, hydronium (or hydroxonium) – which transfers a proton (H^+) to the neutral target molecule (M), depending on the proton affinities (PA) of the two reaction partners (see below), to produce MH^+. Both the SIFT-MS and PTR-MS ion sources also have the capability to produce ions like O_2^+ and NO^+, amongst others, that ionize molecules through charge transfer; this is a characteristic and expedient feature of SIFT-MS (hence "selected ion") (10), but is optionally yet less-commonly implemented in PTR-MS (11). Although they produce a similar range of ions, the ion sources are different in each of the three techniques. All three systems can produce both positively or negatively charged ions, and while positive ions are the most commonly used, negative ions can be useful when analyzing specific chemical classes, such as phenolic compounds, where ionization can be achieved by the removal of a hydrogen atom and the formation of a charged molecule containing an O^- feature. Table 1 summarizes the key features of these technologies, with the details of each discussed in later sections of this chapter.

Table 1. Key features of APCI-MS, PTR-MS and SIFT-MS real-time mass spectrometry systems

Feature	APCI-MS	PTR-MS	SIFT-MS
Ion source	Corona discharge	Hollow cathode	Microwave discharge
Reagent ion	$(H_2O)_nH^+$	H_3O^+ (NO^+, O_2^+)	H_3O^+, NO^+, O_2^+
Ion charge	+ve/-ve	+ve only	+ve/-ve
Buffer gas	Nitrogen	Air (sample matrix)	Helium (optionally N_2)
Sample dilution	Yes	No	Yes
Reaction region	Within corona	Flow drift tube	Fast flow tube
Mass analyzer	Quadrupole, TripleQuad (QqQ)	Quadrupole, Time-of-flight (TOF)	Quadrupole

Returning to the common use of proton transfer reactions using hydronium for CI, and the concept of PA, Figure 1 shows the PA values of different classes of volatile compound; those with a PA value greater than H_3O^+ (165 kcal/mol) will become ionized whereas those below this value will not. Thus, advantageously, common constituents of air – nitrogen, oxygen, argon, carbon dioxide – will not be ionized; accordingly H_3O^+ is widely used as the reagent ion as its PA value is well-placed to ionize the vast majority of volatile compounds of interest while avoiding air constituents. For the NO^+ and O_2^+ reagent ions available in SIFT-MS (12) – and more recently optionally in PTR-MS (11) – the PA concept is not relevant and other mechanisms prevail. The ionization mechanisms involving these reagent ions are well characterized, based on both empirical data and theoretical models (13), and are summarized in Table 2. It should be noted that like H_3O^+, the NO^+ and O_2^+ reagent ions do not ionize the main constituents of air. For O_2^+, ionization of target analytes predominantly proceeds via the electron transfer (ET) mechanism, frequently yielding a parent ion as well as fragment ions. Compared to O_2^+, NO^+ has a somewhat lower ionization energy, which means that fewer volatile compounds react with NO^+ via ET. In many respects, however, this is where the utility of NO^+ in flavor analysis becomes apparent because it provides enhanced selectivity due to different product ion shifts from the ET parent ion. In SIFT-MS, association and hydride abstraction reactions occur with VOCs that do not undergo ET. Typically aldehydes and alcohols react via hydride abstraction, whereas most other carbonyl-containing species (such as ketones, esters, and volatile fatty acids) react via association. For PTR-MS, suprathermal conditions in the flow drift tube (10) prevent association reactions from occurring, reducing the benefit of NO+ to some extent. The benefits of NO+ for flavor applications are discussed in Chapter 4. For the sake of completeness, it should be mentioned that both SIFT-MS and PTR-MS can be operated with alternative reagent ions to the three common ones discussed here. This includes both positive and negative ions, such as NH_4^+ or Kr^+, or O^-, O_2^-, OH^-, NO_2^- and NO_3^-; although each offer benefits for specific applications, their implementation in flavor analysis is limited, thus will not be treated further in this chapter.

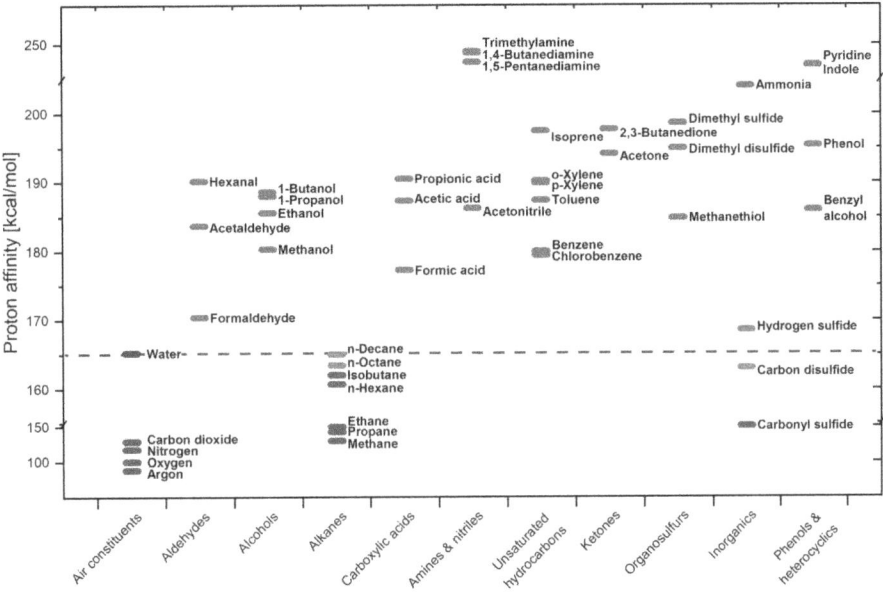

Figure 1. Proton affinity values of common classes of volatile (aroma) compounds. Image used with permission, copyright Fraunhofer IVV, with values taken from (13–15)

Unlike EI-MS, which creates many fragments from a molecule due to its high exothermic nature, CI usually produces the protonated molecular ion (M·H+) along with a few fragments. For maximum sensitivity, concentrating the available charge into a single product ion gives the greatest signal-to-noise (S/N) ratio and therefore the best sensitivity; however, some fragment ions can be useful when trying to identify compounds that produce the same molecular ion. Hexanal and hexenol are good examples of important aroma compounds that are isomeric, i.e., with the same elemental composition ($C_6H_{14}O$) and molecular weight (100.16 g/mol), that cannot be resolved by the M·H+ ion but may be differentiated by the fragment ions, as explained in the section on APCI-MS. The catalog of aroma compounds contains many other examples of isobaric and isomeric molecules, hence the need to use GC-MS to pre-identify the compounds involved in real-time mass spectrometry analyses in order to assign the ions to specific compounds. This deconvolution of the ion profile is another task for the analyst in order to quantify the release of individual compounds (see, for example, (16)). Compounds with the same elemental composition are difficult to distinguish and geometric isomers cannot be resolved by most real-time mass spectrometry techniques.

Table 2. Reaction mechanisms of the three most common reagent ions (H_3O^+, NO^+ and O_2^+) used in the real-time mass spectrometry methods for flavor study applications

Reagent Ion	Ionization criterion	Ionization mechanism(s) (analyte-dependent)	Technique
H_3O^+	Proton affinity* > 165 kcal/mol	Proton transfer (PT) $H_3O^+ + M \rightarrow M \cdot H^+ + H_2O$	APCI-MS, PTR-MS, SIFT-MS
	*(13)	PT is sometimes followed by rapid elimination of H_2O $M \cdot H^+ \rightarrow [M\text{-}OH]^+ + H_2O$	
O_2^+	Ionization energy** < 12.07 eV for ET	Electron transfer (ET) $O_2^+ + M \rightarrow M^+ + O_2$	SIFT-MS (PTR-MS)
		Dissociative ET $O_2^+ + M \rightarrow Fragment^+ + ...$	SIFT-MS (PTR-MS)
NO^+	Ionization energy** < 9.26 eV for ET	Association $NO^+ + M \rightarrow M \cdot NO^+$	SIFT-MS
	**(17)	Hydride abstraction $NO^+ + M \rightarrow [M\text{-}H]^+ + HNO$	SIFT-MS (PTR-MS)
		Also ET and dissociative ET, but less so than O_2^+ (lower ionization potential)	SIFT-MS (PTR-MS)

Comparing the construction of the three technologies, they commonly feature three mechanistic components, namely an ion source, a reaction chamber, and a detection system, as depicted schematically in Figure 2. These components are discussed relevant to each technique in the dedicated sections below. A frequently asked question concerning the different real-time mass spectrometry systems is how their sensitivities compare. This question cannot be answered with a single response, however, as multiple aspects of the analysis must be considered; these include, the nature of the compound, the configuration of the system and the context of the analysis. Compounds with a high PA that produce ions that fall into regions of the spectrum with little background noise, for instance, will be measured at much lower levels compared to compounds with the opposite properties. The type and construction of the mass analyzer used to separate the ions for detection also vary in their transmission efficiencies, which has an impact on detection sensitivity. Generally speaking, all three real-time mass spectrometry systems reviewed here have limits of detection (LODs) in the ppb_V to ppt_V range. In addition to their high sensitivities and low LODs, a particular analytical strength that is common to all three techniques is their broad dynamic range, i.e., their capability to simultaneously detect compounds present at both low and high concentrations. This is especially important in flavor applications, as foods release a complex mixture of aroma compounds over a large quantitative range, and the wide dynamic range of these real-time mass spectrometry technologies means they can detect aroma compounds without concentration-dependent discrimination.

APCI-MS

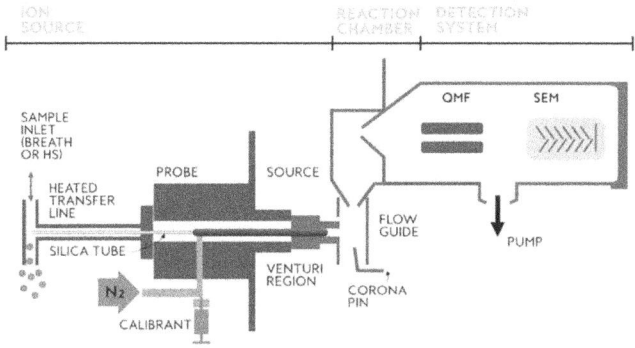

PTR-MS

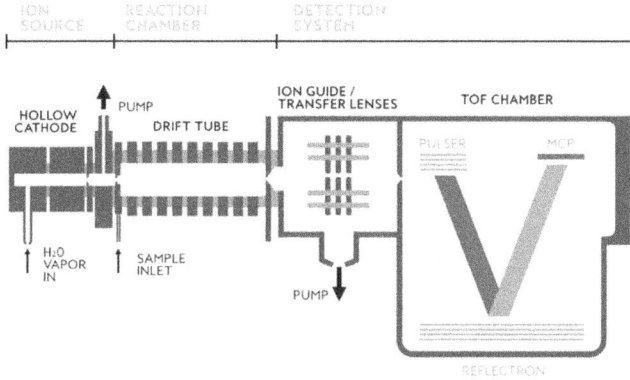

SIFT-MS

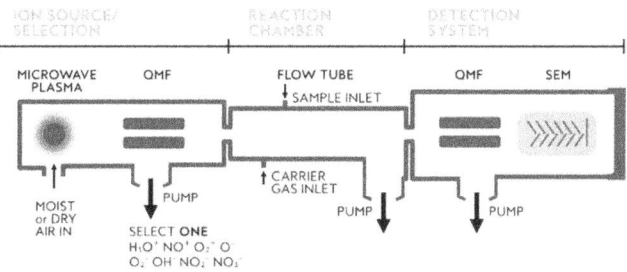

Figure 2. Generic configurations of APCI-MS, PTR-MS and SIFT-MS instrumentation. QMF = quadrupole mass filter; TOF = time-of-flight; SEM = secondary electron multiplier (detector); MCP = multi-channel plate (detector); HS = headspace

The development of APCI-MS, PTR-MS and SIFT-MS systems for dynamic flavor analyses have addressed the issues raised in the previous paragraphs and the way these challenges have been overcome, as well as the different configurations developed (Figure 2), are described in the following sections. It is now possible to obtain reliable data for a wide variety of flavor release applications and the widespread availability of the techniques has led to significant advances in our understanding of aroma release during eating, as well as many other areas, as described in the following chapters in this book.

APCI-MS

In the 1990s, the Flavour Research group at the University of Nottingham, UK was seeking ways to measure aroma release on a breath-by-breath basis, to understand the effect of food matrix composition on the aroma profile delivered to the olfactory receptors during eating. Initially, samples of exhaled air from the nose were trapped on Tenax (2) or short sections of fused silica capillaries cooled in liquid nitrogen (18). The traps or capillaries were then connected to a GC column and the trapped volatiles released and analyzed by conventional GC-MS. This method provided some insight into aroma release *in vivo* but was very time-consuming and prone to significant variation due to the difficulty of capturing each breath on a trap/capillary. A direct injection approach was the obvious answer, but at the time, PTR-MS and SIFT-MS were specialist techniques that were only just emerging and were not readily available. Preliminary experiments with an APCI-MS from Micromass (Manchester, UK; courtesy of Brian Green) showed the potential of this technology for simultaneously monitoring several aroma compounds in real time. APCI-MS was chosen because the equipment was commercially available and the ion source interface, originally designed for use with liquid chromatography, could be easily modified for gas flows.

The challenges of developing APCI-MS for aroma release experiments initially centered around the ion source and the mechanism of ionization via a corona discharge. The corona ionizes water molecules in the incoming gas stream into a mixture of hydronium ions (H_3O^+) and the protonated water dimer $(H_2O)_2H^+$ (19). The proton is then transferred to aroma molecules, giving a protonated molecular ion ($M \cdot H^+$) and some fragments. It was necessary to redesign the original ion source to optimize the process to achieve maximum sensitivity, as well as to deliver a stable platform that gave consistent results. Details of the general APCI-MS process are published (19), as are details of the Nottingham APCI interface (20).

One particular challenge was the necessity to adapt the APCI source so that it could be used to measure aroma release from panels of people. To introduce air samples into the mass spectrometer, a sheathed fused silica capillary was placed inside the buffer gas flow to produce a Venturi effect (Figure 2). Basically, the high buffer-gas flow-rate "sucks" gas through the capillary and the flow rate can be varied by adjusting the position of the capillary using a screw adjuster. A simple solution was adopted to sample exhaled air from panelists using a T-piece fitted with a disposable plastic inlet tube and a heated transfer line between this interface and the connection to the detector. By using heated 530 μm diameter tubing, the high linear flow rate and low dead volume minimized time lag and reduced the degree of condensation of high boiling point volatile compounds. APCI-MS adapts well to the humidity of the sample as the sample flow rate is low – around 10 to 50 mL/min – compared to the higher buffer gas flow rate (2-3 L/min). Therefore, the humidity in the sample accounts for only 0.3 to 2.5% of the humidity in the ion source (on a volumetric basis) and APCI-MS can equally accommodate dry or highly humid samples.

To improve sensitivity of the ion source, as well as assisting in the assignment of ions to compounds present in the sampled air, the cone voltage feature could be exploited. Changing this voltage affects the degree of fragmentation of molecules, and the operating parameters of the Micromass instrument (and the later machines from Waters, Manchester, UK) in selected ion monitoring (SIM) mode could be set to specific *m/z* and cone voltage values, depending on the compounds present in the sample under investigation. In other applications, it was sometimes advantageous to purposely induce fragmentation of certain compounds to differentiate them from compounds that produced the same *m/z* values (see, for example, (21)), as is discussed in more detail in Chapter 2 in relation to the application of APCI-MS to study the volatile release in the headspace above freshly-brewed tea. As mentioned previously, DIMS methods are not designed to identify compounds and it is usual to carry out an initial GC-MS analysis to ascertain which compounds are present in order to set the optimum APCI-MS parameters to monitor these compounds. However, matching ions to compounds is not always straightforward, so a method was developed that utilized GC to separate the compounds followed by simultaneous EI-MS and APCI-MS (22–). Inspection of these ensuing mass spectra allowed better assignment of ions to compounds and it was possible to categorize the confidence of the identification on a scale from "unequivocal" to "tentative" (see Chapter 2 for a practical example).

Calibration of the APCI-MS is carried out either by introducing a solution of the aroma compounds via the calibration port (Figure 2) or by generating headspace of known composition and introducing that via the fused silica capillary line. Because ionization occurs through charge transfer and because the proton affinities of compounds are different (Figure 1), it is possible that high concentrations of a compound with a high PA could "steal" charge from compounds with PAs around that of water and cause ion suppression. This aspect can be monitored by infusing a constant stream of one volatile into the APCI-MS and injecting samples of headspace of another compound to see if the ion signal changes (7).

Although Micromass originally produced a limited number of interfaces to fit the Micromass and Waters ZQ range of mass spectrometers, there was no update for the following Xevo range produced by Waters. Independently of Waters, an interface for the Xevo triple quad MS range (MSNose2) was developed by Jun Hatakeyama at Nichirei Corporation, Japan with design input from Flavometrix Ltd, UK and the triple quad configuration was beneficial in reducing noise and increasing sensitivity over the ZQ single-quad machine (24). Various custom-built APCI-MS systems are still in use around the world, and the 2017 launch of the Advion vAPCI-MS has once again provided a commercial opportunity for aroma release studies using APCI-MS.

PTR-MS

Development of the PTR-MS technique took place concurrently with that of APCI-MS and SIFT-MS. Indeed both PTR-MS and SIFT-MS have the same roots, stemming from the developments of the flowing afterglow technique by Eldon Ferguson, Fred Fehsenfeld, Art Schmeltekopf and colleagues in Boulder, CO in the late 1960s (25). The forerunner to PTR-MS was a SIFT-MS system in which an electric field was applied to the flow tube, to produce a flow-drift tube; this technology was termed selected ion flow drift tube-mass spectrometry (SIFDT-MS). Subsequent work at the Institute of Ion Physics at the University of Innsbruck, Austria leading up to and culminating in the mid-1990s resulted in the transition of the bulky laboratory-based SIFDT-MS system, used primarily for characterizing ion-molecule reactions, into a more compact and portable instrument that was specifically designed for the detection and quantitation of gas-phase VOCs: thus emerged PTR-MS, as it became known (6). Led by Werner Lindinger, Armin Hansel and Alfons Jordan, early work using PTR-MS explored its use in different fields, including flavor research. Recognizing its high potential, Nestlé in Lausanne, Switzerland embraced this emerging technology for aroma studies on coffee, with pioneering work undertaken by Chahan Yeretzian and colleagues that demonstrated the potential of PTR-MS for studying flavor release and firmly positioned it as an emerging technology for dynamic aroma compound analysis (26). Another notable research group that soon adopted PTR-MS in pursuing diverse food/flavor science applications and pushing the boundaries of its potential was the Agri-Food Quality Department at the Istituto Agrario di San Michele all'Adige (now Fondazione Edmund Mach) in Italy. Led by Franco Biasioli, Eugenio Aprea and Flavia Gasperi, and working closely with the University of Innsbruck, the group undertook studies ranging from flavor profiling of mozzarella cheese (27), mass spectral fingerprinting of orange juice (28), and oxidative changes in olive oil (29), to name but a few (detailed further in Chapter 3).

The PTR-MS instrument, like the other two DIMS techniques covered in this chapter, features three sections (Figure 2). Despite its similarities to SIFT-MS, PTR-MS differs in several distinctive aspects. Unlike SIFT-MS, the PTR-MS ion source is a hollow-cathode discharge, whereby water vapor molecules channeled into the source undergo ionization via energetic electrons that generates a dense plasma and an intense cluster of hydronium (H_3O^+) of high purity (>99.5%). Although less common, purging the ion source with N_2 or O_2 instead of water vapor leads to the respective formation of NO^+ or O_2^+ reagent ions, thereby providing switchable reagent ion (SRI) capabilities (11). Interchanging these ions cannot be achieved as rapidly as in SIFT-MS, however, typically in the order of seconds rather than milliseconds. Returning to its primary configuration, i.e., using hydronium, the reagent ions are channeled from the ion source via a Venturi inlet into the reaction chamber of the PTR-MS. As mentioned above, PTR-MS differs from SIFT-MS by the application of an electric field across the flow tube to establish a flow-drift tube. This is the main differentiating feature between the two techniques: where the CI reactions in the SIFT-MS flow tube take place at thermal energy conditions, the drift field in the PTR-MS flow drift tube establishes suprathermal conditions that can be varied by changing the drift voltage (typically between 400-600 V). This has two advantages, yielding more readily interpretable mass spectra: first, the higher energy suppresses water cluster formation, and second, the neutral air (buffer gas) flows through the reaction chamber at a constant rate, whereas the ions (reagents and analytes) are rapidly removed from the drift tube (and into the detection region) by the electric field, thereby reducing the possibility for secondary reactions. As a caveat, the suprathermal conditions result in a slightly higher degree of fragmentation of analyte ions compared to SIFT-MS. A more detailed treatise of the comparability of these two techniques has been published in the literature (30).

At the downstream end of the flow drift tube, the ions are channeled into the detection system. The early (commercial) PTR-MS instrumentation implemented a quadrupole mass filter (QMF) for ion selection, with subsequent detection by a secondary electron multiplier (SEM). More recently, however, time-of-flight (TOF) mass spectrometry has superseded the QMF to become the system of choice for most PTR-MS applications. The first published report of the PTR-MS being coupled to a TOF-MS – generally referred to as PTR-TOF-MS – was in 2004 by Blake and colleagues at the University of Leicester, UK (31), shortly followed by the launch of a commercial system from IONICON Analytik in Innsbruck, Austria (32). In PTR-TOF-MS, the swarm of ions emerging from the drift tube is focused using electrostatic lenses and/or a quadrupole-type ion guide (see Figure 2). This acts to collimate the ions into an intense beam that enters the flight chamber, which reduces losses and increases signal intensities. Upon entering this region, a pulser rapidly accelerates ion bundles orthogonally into the flight chamber, where they encounter a reflectron and are repelled towards the micro-channel plate (MCP) detector that generates an electrical signal for data acquisition. Due to the proportionality of the "time-of-flight" of the ions and the ratio of their mass to charge, i.e., m/z, the latter can be accurately ascertained based on the former.

The advantages afforded by TOF-MS over a QMF are its higher mass resolving power, allowing for many isobaric (but not isomeric) compounds to be distinguished, as well as its speed of analysis, with a full mass scan acquired within milliseconds, and its better sensitivity to higher m/z ions. These features represent tremendous benefits in flavor science, with full mass spectral acquisition at breath-by-breath resolution for *in vivo* studies, as well as high-throughput capabilities for multiple headspace samples of minimal volume. Most PTR-MS systems nowadays in use in flavor research are PTR-

TOF-MS instruments, with commercial instrumentation from IONICON Analytik or Tofwerk in Thun, Switzerland tailored to provide rapid, high resolution and highly sensitive VOC detection at LODs at or below single-digit ppt$_V$ levels (33). Use of PTR-MS for characterizing flavor release has been widespread for the past quarter century and has significantly contributed to our understanding of the related dynamic processes, as are reviewed and further discussed in Chapter 3.

SIFT-MS

David Smith and Nigel Adams invented the selected ion flow tube (SIFT) in 1976 to address shortcomings of earlier laboratory instruments used for studying gas phase ion-molecule reactions in interstellar space and planetary atmospheres (34). By the mid-1990s, several research groups had compiled a database for thousands of these reactions, comprising reaction mechanism details and reaction rate coefficients that were used in computational models. At this time, David Smith and Patrik Spanel realized that by utilizing this database in a new way, SIFT-MS instruments could determine concentrations of compounds directly in the gas phase (35). In the first few years following this landmark publication, Smith, Spanel, and co-workers published numerous articles that characterized reactions of volatile compounds with reagent ions that are relevant to direct analysis of air (H_3O^+, NO^+, and O_2^+) (12) and still support SIFT-MS and PTR-MS applications today. When the first SIFT-MS instruments were developed in the 1990s, LODs were in the mid-parts-per-billion by volume (ppb$_V$) range, but the latest commercial instruments that are applicable for food and flavor research can achieve LODs of low ppt$_V$ (36–).

Spanel and Smith explored food applications with SIFT-MS in 1999 (38), but the literature in this field was sparse until the technique was more recently adopted by several research groups, primarily Frank Devlieghere's group at Ghent University, Belgium (39) and Jim Harper's (37) and Sheryl Barringer's (40) groups at the Ohio State University, OH, USA. Food research using SIFT-MS has since accelerated and the groups publishing in this area have broadened (41).

As with APCI-MS and PTR-MS, there are three main elements of the SIFT-MS technique (Figure 2). In the SIFT-MS ion source, the reagent ions are formed by microwave discharge at low pressure (*ca.* 0.5 Torr) in a moving stream of humid air (i.e., no bottled gas supplies are required). To ensure that just one type of reagent ion is introduced into the reaction chamber, i.e., the fast flow tube, the source gas stream is passed through a QMF. The QMF serves another very important purpose: it enables very rapid switching of reagent ion between H_3O^+, NO^+ and O_2^+ (within tens of milliseconds). This means that selectivity and sensitivity can be optimized in a single analysis by using all three positively charged reagent ions through rapid, sequential filtering through the front-end QMF. Note that utilization of negatively charged reagent ions requires higher pressure conditions in the ion source; since the pressure change takes approximately 30 s, negative and positive reagent ions cannot be used synchronously. To date, negative ion mode has seen little application in food science studies, so is not considered further here, but an example is given and discussed in Chapter 4.

Emerging from the ion source, the selected reagent ion (H_3O^+, NO^+ or O_2^+) is injected into the flow tube reaction chamber and undergoes collisions with the carrier gas (typically helium, although nitrogen is an option on commercial instruments) over a period of about 1 ms, yielding reagent ions of extremely consistent thermal energies. The sample gas is then introduced at a known flow rate (typically 25 mL/min) and the volatile compounds therein are ionized by the reagent ions to form product ions. A further benefit of using a carrier gas is realized: it transports reagent and product ions along the flow tube without the need for adding extra energy via an electric field gradient, thereby minimizing subsequent fragmentation. Together with the collisional cooling of the reagent ions prior to introduction of sample, the result is very consistent ionization. This enables a chemical ionization library to be built in, simplifying method development and data processing.

In the final phase of analysis, product ions and unreacted reagent ions are sampled into the detector system. Here the ions are mass-filtered, then detected using a SEM in pulse-counting mode. By utilizing the integrated compound library, instrument software can instantaneously calculate the concentration of each analyte in real-time in SIM mode. Alternatively, scan mode over a defined m/z range can be employed on static samples (such as headspace) and utilized for rapid volatile "fingerprinting" (see Chapter 4). SIFT-MS technology currently enjoys commercial success, such as the Voice instrument series from Syft Technologies Ltd in Christchurch, New Zealand, in diverse research and industrial applications, including breath analysis in medical studies, food science, and contaminants analysis and monitoring, to name just a few.

Conclusions

Until APCI-MS, PTR-MS and SIFT-MS emerged onto the analytical scene some quarter of a century ago, flavor release studies were accomplished by intermittent sampling and time-consuming GC-MS analyses. The fast analytical capabilities of these three DIMS techniques were quickly recognized and exploited by flavor analysts, with pioneering studies on retronasal aroma perception and persistence via breath-by-breath "nosespace" analysis, and dynamic headspace analysis of flavor partitioning from diverse foods, such as coffee, dairy and meat, as well as in model systems. As the respective technologies became more sophisticated and fine-tuned towards flavor-relevant applications, their scope of implementation expanded, from studying single to multiple compounds simultaneously, with better precision afforded by improved sensitivities and lower LODs that delivered robust, quantitative data with high S/N. Most recently, improvements in data processing software, such as peak deconvolution, feature extraction and automated compound identification, allow for complex, rich datasets to be explored to their fullest extent. Further, the development of new technical features, such as the use of alternative reagent ions, or the replacement of the conventional QMF with QqQ or TOF mass spectrometers, offer yet further analytical power.

Nowadays, real-time mass spectrometry represents an indispensable tool for exploring dynamic flavor release. This introductory chapter has sought to offer the reader an overview of APCI-MS, PTR-MS and SIFT-MS, presenting their common and differing operating principles – with a brief discourse on their early developments – and highlighting their versatility and strengths for studying aroma compounds under changing conditions. The ensuing chapters of this book provide specific examples of their application to, and relative merits for, the study of flavor release in real-time, with reviews of past, key studies and previously unpublished research, that clearly demonstrate that there are yet more breakthroughs to be achieved in flavor science using these sophisticated tools.

Conflict of Interest

AJT and JDB declare no conflict of interest. VSL is principal scientist at Syft Technologies Limited, a manufacturer of SIFT-MS instrumentation.

Abbreviation List

APCI-MS	atmospheric pressure chemical ionization-mass spectrometry
CI	chemical ionization
DIMS	direct injection-mass spectrometry

EI	electron ionization
ET	electron transfer
GC-MS	gas chromatography-mass spectrometry
LOD	limit of detection
MCP	multi-channel plate (detector)
MS	mass spectrometry
m/z	mass-to-charge (ratio)
PA	proton affinity
PT	proton transfer
PTR-MS	proton transfer reaction-mass spectrometry
PTR-TOFMS	proton transfer reaction-time-of-flight-mass spectrometry
QMF	quadrupole mass filter
QqQ	triple quadrupole (MS)
SEM	secondary electron multiplier (detector)
SIFDT-MS	selected ion flow drift tube-mass spectrometry
SIFT-MS	selected ion flow tube-mass spectrometry
SIM	selected ion monitoring (measurement mode)
S/N	signal-to-noise (ratio)
SRI	switchable reagent ion
TOF	time-of-flight (MS)
VOC	volatile organic compound

References

1. Buettner, A.; Beauchamp, J. Chemical input - sensory output: diverse modes of physiology-flavour interaction. *Food Qual. Prefer.* **2010**, *21*, 915–924.
2. Ingham, K. E.; Linforth, R. S. T.; Taylor, A. J. The effect of eating on aroma release from strawberries. *Food Chem.* **1995**, *54*, 283–288.
3. Benoit, F. M.; Davidson, W. R.; Lovett, A. M.; Nacson, S.; Ngo, A. Breath analysis by atmospheric-pressure ionization mass-spectrometry. *Anal. Chem.* **1983**, *55*, 805–807.
4. Soeting, W. J.; Heidema, J. A mass spectrometric method for measuring flavour concentration/time profiles in human breath. *Chem. Senses* **1988**, *13*, 607–617.
5. Smith, D.; Spanel, P. The novel selected-ion flow tube approach to trace gas analysis of air and breath. *Rapid Commun. Mass Spectrom.* **1996**, *10*, 1183–1198.
6. Hansel, A.; Jordan, A.; Holzinger, R.; Prazeller, P.; Vogel, W.; Lindinger, W. Proton transfer reaction mass spectrometry: on-line trace gas analysis at the ppb level. *Int. J. Mass Spectrom. Ion Proc.* **1995**, *149–150*, 609–619.
7. Taylor, A. J.; Linforth, R. S. T. In *Food Flavour Technology*, 2nd ed.; Taylor, A. J., Linforth, R. S. T., Eds.; Wiley-Blackwell: Chichester, 2010; pp 266–295.
8. Munson, M. S. B.; Field, F. H. Chemical ionization mass spectrometry. I. General Introduction. *J. Am. Chem. Soc.* **1966**, *88*, 2621–2630.
9. Ferguson, E. E. A personal history of the early development of the flowing afterglow technique for ion-molecule reaction studies. *J. Am. Soc. Mass Spectr.* **1992**, *3*, 479–486.

10. Smith, D.; McEwan, M. J.; Španěl, P. Understanding gas phase ion chemistry is the key to reliable selected ion flow tube-mass spectrometry analyses. *Anal. Chem.* **2020**, *92*, 12750–12762.
11. Jordan, A.; Haidacher, S.; Hanel, G.; Hartungen, E.; Herbig, J.; Märk, L.; Schottkowsky, R.; Seehauser, H.; Sulzer, P.; Märk, T. D. An online ultra-high sensitivity proton-transfer-reaction mass-spectrometer combined with switchable reagent ion capability (PTR+SRI-MS). *Int. J. Mass Spectrom.* **2009**, *286*, 32–38.
12. Smith, D.; Španěl, P. Selected ion flow tube mass spectrometry (SIFT-MS) for on-line trace gas analysis. *Mass Spectrom. Rev.* **2005**, *24*, 661–700.
13. Hunter, E. P.; Lias, S. G. Evaluated gas phase basicities and proton affinities of molecules: an update. *J. Phys. Chem. Ref. Data* **1998**, *27*, 413–656.
14. Blake, R. S.; Patel, M.; Monks, P. S.; Ellis, A. M.; Inomata, S.; Tanimoto, H. Aldehyde and ketone discrimination and quantification using two-stage proton transfer reaction mass spectrometry. *Int. J. Mass Spectrom.* **2008**, *278*, 15–19.
15. Hunter, K. C.; East, A. L. L. Properties of C−C bonds in n-alkanes: relevance to cracking mechanisms. *J. Phys. Chem. A* **2002**, *106*, 1346–1356.
16. Wright, J.; Wulfert, F.; Hort, J.; Taylor, A. J. Effect of preparation conditions on release of selected volatiles in tea headspace. *J. Agric. Food Chem.* **2007**, *55*, 1445–1453.
17. Lias, S. G. Ionization Energy Evaluation. NIST Standard Reference Database Number 69. In *NIST Chemistry WebBook*; Linstrom, P. J.; Mallard, W. G.; National Institute of Standards and Technology (accessed May 19, 2021).
18. Ingham, K. E.; Linforth, R. S. T.; Taylor, A. J. The effect of eating on the rate of aroma release from mint-flavored sweets. *Lebensmittel Wissenschaft Technologie* **1995**, *28*, 105–110.
19. Byrdwell, W. C. Atmospheric pressure chemical ionization mass spectrometry for analysis of lipids. *Lipids* **2001**, *36*, 327–346.
20. Linforth, R. S. T.; Taylor, A. J. Apparatus and methods for the analysis of trace constituents of gases. EP 0819 937 A2, 1998.
21. Jublot, L.; Linforth, R. S. T.; Taylor, A. J. Direct Atmospheric Pressure Chemical Ionisation Ion Trap Mass Spectrometry for Aroma Analysis: Speed, Sensitivity and Resolution of Isobaric Compounds. *Int. J. Mass Spec.* **2005**, *243*, 269–277.
22. Taylor, A. J.; Sivasundaram, L.; Linforth, R. S. T.; Surawang, S. Identification of volatile compounds using combined API/EI-MS; In *Abstracts of Papers of the American Chemical Society*; Aug. 18, 2002; 033-AGFD.
23. Taylor, A. J.; Sivasundaram, L. R.; Linforth, R. S. T.; Surawang, S. In *Handbook of Flavor Characterization. Sensory Analysis, Chemistry and Physiology*; Deibler, K. D., Delwiche, J., Eds.; Marcel Dekker: New York, 2003; pp 411−422.
24. Hatakeyama, J.; Taylor, A. J. Optimization of atmospheric pressure chemical ionization triple quadropole mass spectrometry (MS Nose 2) for the rapid measurement of aroma release in vivo. *Flavour Fragr. J.* **2019**, *34*, 307–315.
25. Ferguson, E. E.; Fehsenfeld, F. C.; Schmeltekopf, A. L. In *Advances in Atomic and Molecular Physics*; Bates, D. R., Estermann, I., Eds.; Academic: New York, 1969; Vol. 5, pp 1–56.

26. Yeretzian, C.; Jordan, A.; Badoud, R.; Lindinger, W. From the green bean to the cup of coffee: investigating coffee roasting by on-line monitoring of volatiles. *Eur. Food. Res. Technol.* **2002**, *214*, 92–104.
27. Gasperi, F.; Gallerani, G.; Boschetti, A.; Biasioli, F.; Monetti, A.; Boscaini, E.; Jordan, A.; Lindinger, W.; Iannotta, S. The mozzarella chesse flavour profile: a comparison between judge panel analysis and proton transfer reaction mass spectrometry. *J. Sci. Food Agric.* **2000**, *81*, 357–363.
28. Biasioli, F.; Gasperi, F.; Aprea, E.; Colato, L.; Boscaini, E.; Märk, T. D. Fingerprinting mass spectrometry by PTR-MS: heat treatment vs. pressure treatment of red orange juice--a case study. *Int. J. Mass Spectrom.* **2003**, *223–224*, 343–353.
29. Aprea, E.; Biasioli, F.; Sani, G.; Cantini, C.; Mark, T. D.; Gasperi, F. Proton Transfer Reaction−Mass Spectrometry (PTR-MS) Headspace Analysis for Rapid Detection of Oxidative Alteration of Olive Oil. *J. Agric. Food Chem.* **2006**, *54*, 7635–7640.
30. Smith, D.; Španěl, P.; Herbig, J.; Beauchamp, J. Mass spectrometry for real-time quantitative breath analysis. *J. Breath Res.* **2014**, *8*, 027101.
31. Blake, R. S.; Whyte, C.; Hughes, C. O.; Ellis, A. M.; Monks, P. S. Demonstration of proton-transfer reaction time-of-flight mass spectrometry for real-time analysis of trace volatile organic compounds. *Anal. Chem.* **2004**, *76*, 3841–3845.
32. Jordan, A.; Haidacher, S.; Hanel, G.; Hartungen, E.; Märk, L.; Seehauser, H.; Schottkowsky, R.; Sulzer, P.; Märk, T. D. A high resolution and high sensitivity proton-transfer-reaction time-of-flight mass spectrometer (PTR-TOF-MS). *Int. J. Mass Spectrom.* **2009**, *286*, 122–128.
33. Sulzer, P.; Hartungen, E.; Hanel, G.; Feil, S.; Winkler, K.; Mutschlechner, P.; Haidacher, S.; Schottkowsky, R.; Gunsch, D.; Seehauser, H.; Striednig, M.; Jürschik, S.; Breiev, K.; Lanza, M.; Herbig, J.; Märk, L.; Märk, T. D.; Jordan, A. A proton transfer reaction-quadrupole interface time-of-flight mass spectrometer (PTR-QiTOF): high speed due to extreme sensitivity. *Int. J. Mass Spectrom.* **2014**, *368*, 1–5.
34. Adams, N. G.; Smith, D. The selected ion flow tube (SIFT); A technique for studying ion-neutral reactions. *Int. J. Mass Spectrom. Ion Phys.* **1976**, *21*, 349–359.
35. Smith, D.; Španěl, P. The novel selected-ion flow tube approach to trace gas analysis of air and breath. *Rapid Commun. Mass Spectrom.* **1996**, *10*, 1183–1198.
36. Langford, V. S.; McEwan, M. J.; Perkins, M. J. High-throughput analysis of volatile compounds in air, water, and soil using SIFT-MS. *Curr. Trends Mass Spectrom.* **2018**, *37*, 24–29.
37. Harper, W. J.; Kocaoglu-Vurma, N. A.; Wick, C.; Elekes, K.; Langford, V. In *Volatile Sulfur Compounds in Food*; American Chemical Society: 2011; Vol. 1068, pp 153–181.
38. Španěl, P.; Smith, D. Selected ion flow tube - mass spectrometry: detection and real-time monitoring of flavours released by food products. *Rapid Commun. Mass Spectrom.* **1999**, *13*, 585–596.
39. Noseda, B.; Dewulf, J.; Goethals, J.; Ragaert, P.; Van Bree, I.; Pauwels, D.; Van Langenhove, H.; Devlieghere, F. Effect of food matrix and pH on the volatilization of bases (TVB) in packed North Atlantic gray shrimp (Crangon crangon): volatile bases in MAP fishery products. *J. Agric. Food Chem.* **2010**, *58*, 11864–11869.

40. Xu, Y.; Barringer, S. Effect of temperature on lipid-related volatile production in tomato puree. *J. Agric. Food Chem.* **2009**, *57*, 9108–9113.
41. Langford, V. S.; Padayachee, D.; McEwan, M. J.; Barringer, S. A. Comprehensive odorant analysis for on-line applications using selected ion flow tube mass spectrometry (SIFT-MS). *Flavour Frag. J.* **2019**, *34*, 393–410.

Chapter 2

Flavor Applications of Direct APCI-MS

Andrew J. Taylor

Flavometrix Limited, Loughborough, Leicestershire, United Kingdom
Emeritus Professor, University of Nottingham, United Kingdom
*Email: flavometrix@btconnect.com

Direct atmospheric pressure chemical ionization-mass spectrometry (APCI-MS) was originally modified by the Nottingham flavor research group to measure aroma release from humans during eating, with the aim of defining the aroma profile that was delivered to the olfactory receptors. The concept was that this type of analysis would provide qualitative and quantitative data to help in formulating successful flavorings for different food matrices. Besides the mass spectrometer modifications, it was also necessary to find ways to interrogate the data, identify compounds from the APCI mass spectra and develop ways to process the data to address issues like human variation. This chapter outlines the applications of direct APCI-MS to aroma release *in vivo*, and then describes how the capability to rapidly measure volatile compounds, led to a variety of other analytical flavor applications.

Learnings from *In Vivo* Aroma Release Analyses

In the 1990s, food manufacturers in many countries were being challenged to produce products with decreased amounts of fat to meet governmental nutritional guidelines. It became clear that the flavor of many low-fat foods was unacceptable and that flavor formulations needed to be revised to restore the desired flavor characteristics (*1*). As proof of principle that direct atmospheric pressure chemical ionization-mass spectrometry (APCI-MS) could be a useful tool in reformulating flavors through design (rather than through trial and error), a series of experiments were carried out at the University of Nottingham (Nottingham, UK) using yogurt as the food matrix. Yogurts with different levels of fat were prepared with aroma compounds that covered a wide range of hydrophobicity, from water-soluble compounds, through to highly hydrophobic, fat-soluble compounds (*2*). The yogurt flavor was assessed using sensory protocols (including time-intensity), as well as by direct APCI-MS, to measure aroma release on a breath-by-breath basis. Results showed that aroma release profiles in products with fat levels between 3 and 10% were almost identical but, below 3% fat, the hydrophobic compounds were over-expressed in the aroma profile. Moreover, it was possible to estimate the degree of over-expression and reformulate the flavoring to try and match the desired "full-fat" release profile, as well as indicating fat levels where the use of the standard flavor formulation was valid. The experiments also gave credence to the concept of fat acting as an aroma "reservoir" and, some years later, ß-cyclodextrin was used as a model "flavor reservoir" in low fat yogurts (3). Compared to standard low-fat yogurts, the cyclodextrin-containing yogurts were preferred as they delivered both the desired level of aroma compound and also better mimicked the temporal delivery of aroma (3). The Nottingham studies of aroma release during eating, using APCI-MS, covered many different types of foods and beverages (summarized in Table 1).

Table 1. Publications on aroma release from different food types measured by APCI-MS at the University of Nottingham, UK

Commodity	References
Emulsions	(4–7)
Thickened systems	(8, 9)
Custard	(10, 11)
Yogurt	(2, 3, 12)
Cheese	(13, 14)
Biscuits	(15–17)
Confectionery	(18–27)
Tomatoes	(12, 28, 29)
Kiwifruit	(30)
Beverages	(31–41)

Aroma and taste release experiments at Nottingham also identified a strong cross-modal interaction between sweetness and mint flavor in chewing gum (42). Sensory assessments of mintiness were found to follow the release of sugar rather than the release of mint aroma compounds. Therefore, control of the sweetness stimulus (whether via sugar or high-intensity sweeteners) could prolong the sensation of mintiness as evidenced by the development of Stride chewing gum. This product was marketed in the USA with the strap-line "the gum with the ridiculously long-lasting flavor" and used encapsulated sweeteners to prolong the perception of mintiness through a cross-modal interaction. To further study cross-modal interactions in the laboratory, it was necessary to deliver aroma and taste compounds to panelists via a gustometer, while, at the same time, monitoring the levels of taste and aroma compounds in the mouth and nose, as well as recording sensory perception through time-intensity measurements (43, 44). Data from these experiments identified several other cross-modal interactions, such as sugar/acid/aroma (25), savory flavors (45) and sub-threshold taste stimuli (46), effects that were observed in trained panelists as well as in naïve subjects.

With strong evidence for interactions between taste and aroma, attention turned to the potential of texture-taste-aroma interactions. Besides satisfying academic curiosity, the results also had relevance for the food industry where thickeners are widely used and often substituted to keep costs down when one class of thickener becomes expensive, due to availability or climatic reasons. Working with this three-component system made the design of the experiments more difficult, but the conclusion was that some food thickeners did change the perceived flavor. However, the aroma release was not significantly changed; instead it was postulated that the release of tastants was the driving factor for the change in savory and sweet flavor perception (*8*, *47–49*). This latter observation was supported by later work from Brossard *et al.* (*50*). Several other research groups have also used *in vivo* release in combination with other analytical techniques to study cross/multimodal interactions, see for example (*13*, *51–56*).

The development of commercially-relevant analytical techniques led to the establishment of Flavometrix in 2002 as a spin-out company from the University of Nottingham, where it set up self-contained laboratories and provided support to many large and small food and flavor companies up to 2014. With the availability of commercial proton transfer reaction-mass spectrometry (PTR-MS) and selected ion flow tube-mass spectrometry (SIFT-MS) systems, aroma release measurements have become more commonplace in both research and commercial laboratories, with strong research centers in Europe (e.g., in Germany (*57*), Italy (*58*), France (*59*)) and the USA (e.g., Ohio (*60*, *61*)). These studies have significantly increased our understanding of the aroma release process and its effect on flavor perception, which has led to the development of smarter flavorings for commercial use.

Dealing with Data from Direct APCI-MS

Analysis of Aroma Compounds during Brewing of Tea

Hot beverages like tea and coffee deliver significant sensory appeal before drinking by releasing aromas from the cup to the nose via the orthonasal route (sniffing through the nostrils), as well as during and after drinking through the retronasal route (when aromas move from the mouth through the throat to the olfactory receptors in the nose). Measuring the orthonasal aroma signal as tealeaves are "brewed" in the cup is important for tea manufacturers; one reason is the ability to monitor the consistency of tea blends over time and the other is to identify differences between teas of different origin. The atmosphere above brewing tea is very humid and also very dynamic and inconsistent. Temperature variations between the hot tea in a cup and the surrounding air causes localized air currents to form, while air movements in the room can move and dilute the "plume" of aroma release from the cup.

Preliminary experiments showed that the variation in air flow above a cup of tea containing boiled water was so strong that reproducible analysis of aroma release could not be obtained. Instead, airflow around the cup was semi-controlled by placing a cover over the cup (Figure 1) and encouraging air flow to enter at the bottom and exit at the top (*39*). Sampling inside the cover then delivered a more reproducible headspace concentration with variation around 5% (*n*=25).

Having found a solution to sampling the odors above brewing tea, the next issue was to identify the released compounds from their ions in the APCI ion trace. To tackle this challenge, the headspace from brewing tea was trapped on a solid-phase microextraction (SPME) fiber and analyzed by gas chromatography (GC) with simultaneous mass spectrometric detection via electron ionization (EI) (to identify the compounds) and APCI (*39*). Ions detected in the APCI trace of the chromatogram could then be matched to compounds identified by EI according to their retention times. For each compound, the expected APCI ions were monitored across the chromatogram using selected ion mode and the retention times at which they appeared were noted (Figure 2).

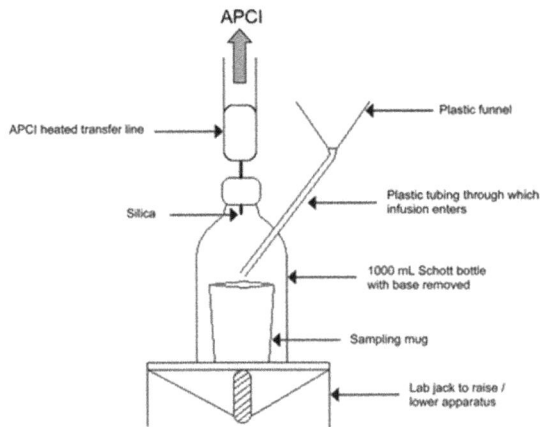

Figure 1. Schematic of a system for monitoring volatile release from tea during brewing. Reproduced with permission from reference (39). Copyright 2007 ACS

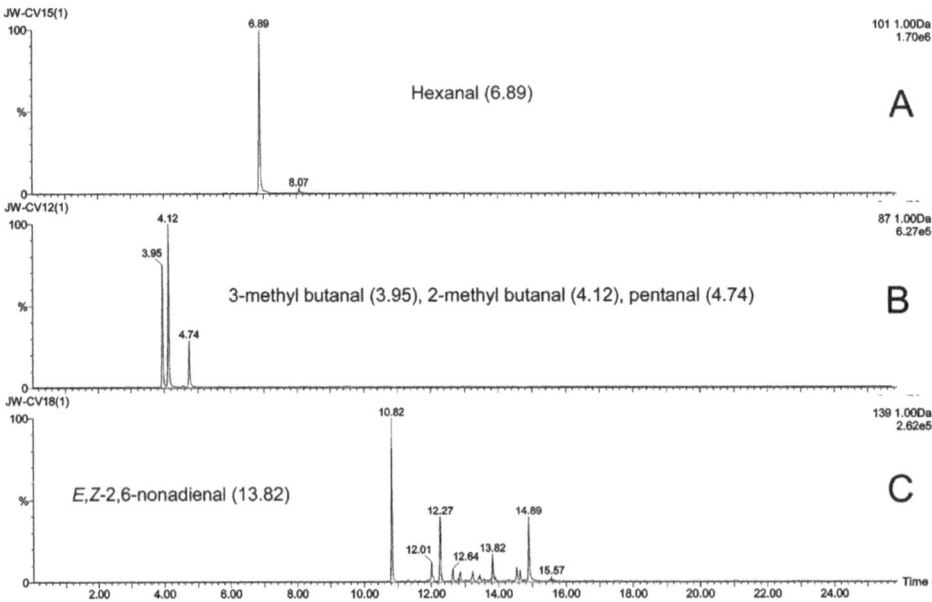

Figure 2. GC-APCI-MS selected ion traces from the analysis of a SPME extract of tea headspace. Trace A shows that only two compounds in the extract have an APCI ion at m/z 101 and that hexanal (retention time 6.89 min) accounts for the majority of the signal. Traces B and C show that ions at m/z 87 and 139 are associated with three or more compounds respectively. Reproduced with permission from reference (39). Copyright 2007 ACS.

Three types of matches were observed: ion traces that clearly matched a single compound, ion traces that could be attributed to several known compounds, or ion traces that related to multiple compounds, including identified compounds. To illustrate this, the compound hexanal (trace A in Figure 2) has an expected APCI ion [MH]$^+$ at m/z 101 and a single major peak was observed at 6.89 min with a much smaller peak at 8.07 min. From the EI trace, we know that hexanal elutes at 6.89 min, so the hexanal peak accounts for the vast majority of ions at m/z 101. In trace B, m/z 87 is the selected ion and this corresponds with the elution of three compounds (2- and 3-methylbutanal and pentanal). In trace C, the ion at m/z 139 is associated with multiple compounds. Therefore, calculating the exact amounts of each of these compounds is not possible from traces B and C in Figure 2.

The inability to assign ions to compounds in complex flavor mixtures when working with real foods is a limitation of all the on-line direct MS techniques. Some leverage can be gained by preparing a food with a "flavoring" that consists of volatile compounds with molecular weights (and corresponding ions) that do not interfere with each other and which represent a spread of hydrophobicity. By measuring the release of this "model flavor" in human panelists, it is possible to predict the release of any compound from a given food, based on its physicochemical properties (62, 63). However, in the case of food products like tea, the natural flavors are not free but embedded in a biological matrix and adding "model flavors" as proposed in the previous sentence, will not match the release characteristics of the embedded compounds.

Dealing with Aroma Release Variation in Human Panelists

Human variation is a well-established factor in sensory evaluation and is also observed during *in vivo* aroma release experiments. The way that humans chew and swallow food (known as oral processing) depends on a variety of factors, such as the panelists' dental status, the frequency of chewing and swallowing, as well as saliva flow rates (64). To reduce variation, some aroma release experiments have been carried out using a fixed chewing protocol, for example, using a metronome to synchronize chewing rates across a sensory panel. Although this controls one factor, it still leaves many others uncontrolled. There are always conflicts when considering the design of experiments that involve humans. If human behaviors are uncontrolled, it may result in datasets with variations that are so large that statistical analysis cannot produce meaningful interpretation of the results. On the other hand, imposing an artificial eating protocol on a panel may give more consistent results but does not give a true picture of what actually happens in real-life, consumer situations.

To measure the level of variation in humans, an "uncontrolled" experiment was devised with a consumer panel ($n=98$) (65). Participants were asked to assess two samples containing the same amount of ethyl hexanoate (347 mg/L) in either regular milk (3.6 g/L fat) or in a low-fat milk (0.1 g/L fat). The samples were rated by panelists to establish which sample had the highest "fruity" note; simultaneously, exhaled air from their noses was monitored using APCI-MS, referred to as nosespace analysis. Sensory results indicated that panelists recorded a significantly higher fruity intensity ($p<0.0001$) in the low fat milk compared to the regular fat milk. For each panelist, the APCI data were transformed by calculating the ratio of the maximum ion intensity for ethyl hexanoate in nosespace from the two milk samples. By ratioing the results, rather than quoting absolute or overall mean values, some of the "noise" in the dataset was decreased. This ratio was termed the "Lipid Effect" (LE) and was further transformed to a logarithmic scale to obtain a Poisson distribution and make the data suitable for parametric statistical analysis (65).

Figure 3A shows a plot of log(LE) versus the number of panelists recording that LE value. A log(LE) value of 0 indicates an equal release of ethyl hexanoate in the two milk samples, but the values for the low fat milk are centered at a log(LE) of about 0.2, showing that ethyl hexanoate release was higher in the low fat samples. Over all the panelists, the variation in the ratio of release values was around 10 times. A few panelists recorded values of log(LE)<0, suggesting they released more ethyl hexanoate from the regular fat sample compared to the low fat milk sample, but this is more likely due to errors in oral processing and/or exhaling air into the APCI-MS inconsistently during the experiment. The mean ratio value from Figure 3A was then taken and used to calculate the theoretical concentration of ethyl hexanoate required to give the same release in the low and regular fat samples, then the sensory and APCI analyses were repeated (65). The data in Figure 3B shows that the theoretical adjustment to aroma levels re-centers the frequency plot at log(LE) 0, the value that indicates the same aroma release in both samples. Therefore, using a consumer panel can deliver meaningful results in this simple study if the data are treated by ratioing the effect in each panelist, rather than trying to take a mean of the absolute release values (e.g., ion intensity).

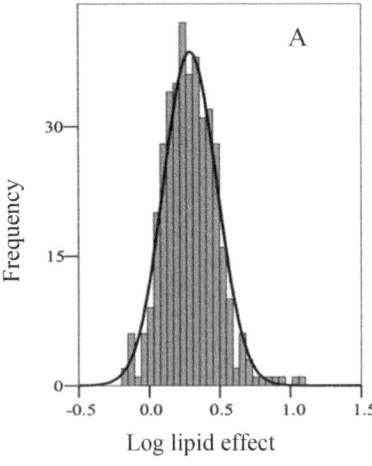
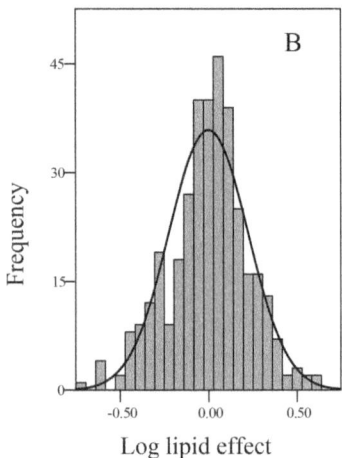

Figure 3. A: Frequency distribution of the log lipid effect for regular and low fat milk samples each flavored with ethyl hexanoate (347 mg/L). B: Frequency distribution after adjusting the concentration of ethyl hexanoate in low fat milk to theoretically match release in both milk samples. Four replicates were carried out with the panel and each distribution is based on ~390 values. Reproduced with permission from reference (65). Copyright 2006 Wiley.

Understanding Dynamic Flavor Systems Using Direct APCI-MS

Dynamic Headspace Dilution Analyses and the Effect of Ethanol

As discussed in Chapter 1, equilibrium headspace measurements have limited relevance to aroma release situations in real life. To mimic aroma release that is sensed by the orthonasal pathway, dynamic headspace dilution (DHSD) techniques give valuable information about the effects of the liquid phase composition on release, whether that be a solute like sucrose or a co-solvent like ethanol. Typically, a DHSD experiment involves placing a known volume of liquid into a container with a known headspace (HS) volume and leaving the system to attain equilibrium before introducing a flow of gas (usually air or nitrogen) across the HS and monitoring the change in concentration with time. Under these highly controlled conditions, a combination of physical chemistry and mass-transfer principles can be used to predict aroma release and compare it with real experimental data (66). DHSD has been applied in a range of food systems, such as the potential for dissolved sugar to change the partition of aroma compounds between the liquid and air phases (67), as well as the aroma release rates from food emulsions (6). The effect of ethanol on aroma release is an important factor in the enjoyment of alcoholic beverages, since aroma release in the glass before drinking delivers the orthonasal "bouquet" of a wine or the "nose" of spirits like whisky and brandy. Although ethanol-water mixtures appear to be a simple binary system, it is known that the mixtures are non-ideal due to the heterogeneous distribution of ethanol depending on the ethanol/water ratio (68). Given the importance of the orthonasal signal in alcoholic spirits, there is a desire to understand the mechanisms involved in the aroma release process and DHSD is a good tool to measure release under different conditions, such as alcoholic strength or the presence of multiple aroma compounds. The HS above ethanolic beverages contains ethanol at high levels and aroma compounds at significantly lower levels, a situation that interferes with ionization and can cause non-quantitative results. In the Nottingham APCI-MS, a constant stream of ethanol was introduced into the ion source so that protonated ethanol acted as the proton transfer agent (*41*) and the aroma compounds could be measured quantitatively. This technique was later applied to analyze alcoholic systems using PTR-MS (see Chapter 3).

Some simple studies comparing the aroma release of different compounds from a 12% solution of ethanol/water (representing wine) and an aqueous solution, indicated why reducing alcohol content, while maintaining the orthonasal signature, is so difficult. Figure 4 shows the release of four aroma compounds (with different hydrophobicity values) from an aqueous or a 12% ethanol/water solution. Initially the equilibrium HS values are higher for the aqueous solution. This can be explained because ethanol has a greater solvating power for the aroma compounds and they are better retained in the liquid phase. However, as the headspace is diluted with air, the concentration above the aqueous solution drops rapidly while the release from the ethanolic solution is affected to a much lesser degree.

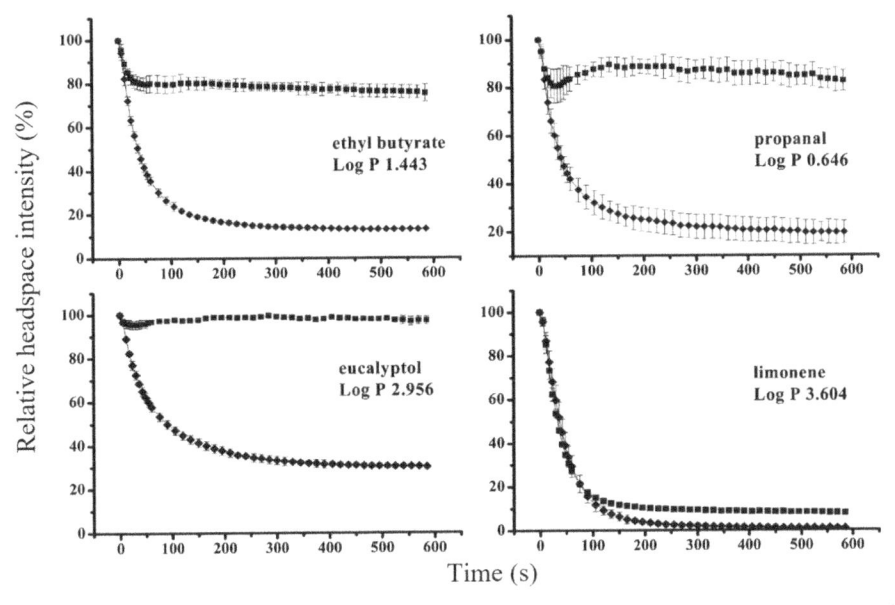

Figure 4. Plot of headspace ion intensity against time of headspace dilution with air for aqueous solutions (diamonds) and 12% ethanol/water (squares) containing four volatiles. Reproduced with permission from reference (36). Copyright 2008 ACS.

With the benefit of hindsight and more experience working with very hydrophobic compounds in ethanol/water mixtures, the limonene data is probably skewed, due to some undissolved limonene that locates itself at the air-water interface and over-estimates the equilibrium headspace concentration. Ensuring that poorly-soluble compounds are actually solubilized in water or ethanol/water systems (and not dispersed or congregated at the air-water interface) is a significant experimental problem in studies like these.

The presence of ethanol in aqueous solution causes changes to the surface tension of the solution due to ethanol molecules congregating at the air-water interface. Dilution of the headspace with gas results in the evaporation of some ethanol, causing cooling of the surface layer. The cooled liquid then sinks through the liquid and ultimately sets up convection currents in the bulk phase, the so-called Marangoni effect, which effectively stirs the liquid phase and significantly changes the mass transfer properties of the system (69, 70). The speed of the direct analysis allows numerous samples to be analyzed, therefore, the effect of many different formulations on aroma release can be tested to build a comprehensive picture of the influence of different alcohol concentrations (37), different co-solutes or other surface-active agents (34).

Maillard Reaction Mechanisms

Generation of aroma and taste compounds through the Maillard reaction is a long-standing topic of interest for flavor scientists. The key stages of the reaction were identified in 1953 by Hodge (71) and subsequent publications have constructed detailed pathways for the different stages. Figure 5 shows the first stage of the reaction mechanism between valine with the dicarbonyl compound pyruvaldehyde. The pathway involves an addition reaction, followed by loss of a water molecule (dehydration), loss of carbon dioxide (decarboxylation), then addition of a water molecule (hydrolysis) to form two smaller compounds. One of the unknown factors is how water content affects the dehydration/hydrolysis stages.

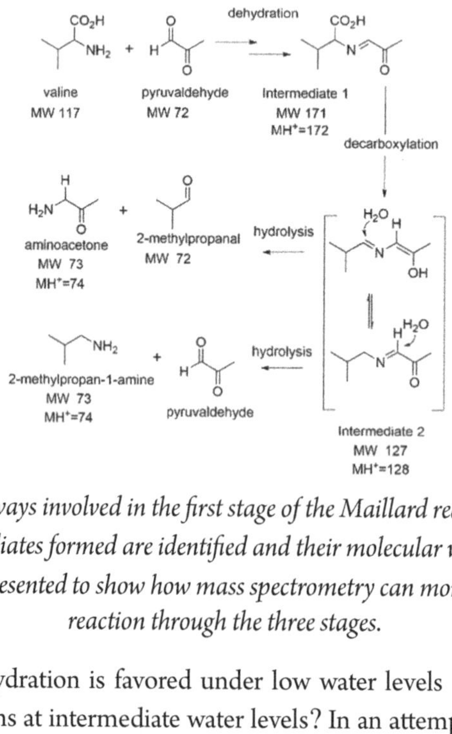

Figure 5. Chemical pathways involved in the first stage of the Maillard reaction between valine and pyruvaldehyde. The intermediates formed are identified and their molecular weight (MW) and protonated m/z values (MH+) are presented to show how mass spectrometry can monitor the progression of the reaction through the three stages.

Theory states that dehydration is favored under low water levels and hydrolysis is favored at high levels, but what happens at intermediate water levels? In an attempt to answer this question, a three-stage system was built (72). The first stage created a humidified air flow (at known levels and at defined temperatures). The second stage contained the valine and pyruvaldehyde reactants, which were initially applied to a bed of inert glass beads as a solution, then dried by passing warm air over the bed. The third stage was APCI-MS detection of the intermediate compounds. From Figure 5, it can be seen that the dehydrated addition reaction is characterized by the production of a compound with MW 171, which becomes protonated to give an ion at m/z 172. Decarboxylation produces two isomeric compounds with MW 127 (m/z 128); hydrolysis of these compounds results in a mixture of four compounds, two with the same MW as pyruvaldehyde (one of the starting compounds) and two with MW values of 73 (m/z 74). Therefore, by monitoring the ion signals at m/z values of 172, 128 and 74 as the reaction mixture was heated, it was possible to follow the course of the reaction (addition/dehydration, decarboxylation and hydrolysis) as humidity levels were changed. When the temperature of the system was held at 170°C, and at intermediate humidity, the dominant ion was m/z 128, suggesting that this humidity level allowed some dehydration to occur in the first step of the reaction mechanism but that the level was too low to favor hydrolysis in the third step. Changing the humidity to zero resulted in m/z 172 becoming the major ion in the spectrum, indicating that dehydration was favored more at this humidity level but that the subsequent steps in the reaction were not favored. A high humidity level led to a brief increase in m/z 74, with a complete loss of signal at m/z 172 and 128, suggesting that only the final step in the reaction sequence was favored at this level. Although the system described above is complex to build and operate, the data it delivers provides unique information on the way the initial stages of the Maillard reaction proceed and the effect of temperature and humidity levels on reaction rates. Further details of the procedures and the results are available in the literature (72–74).

Insect Behavior Studies

Direct APCI-MS was developed at the rural Sutton Bonington campus of the University of Nottingham where there was an emphasis on sharing research outcomes with colleagues in different areas of biological sciences. One such conversation with Ian Hardy's team resulted in a challenge as to whether it was possible for direct APCI-MS to monitor the emission of volatile compounds when insects fought each other. An initial experiment captured the released chemical compound from small wasps and GC-MS analysis identified it as a known compound in the Bethylinae species, 2-methyl-1,7-dioxaspiro[5.5]undecane (75). Next the system for monitoring the insect contests was adapted to allow air sampling from the contest chamber into the APCI-MS and the lag time from contest to APCI detection was established. When two wasps challenged each other, the APCI could detect the timing and quantity of release, but whether the winner or loser of the contest had emitted the compound was not conclusive. The solution was to feed one set of wasps with deuterated water, so as to label the spiro compound and mark the two groups with a tiny dot of acrylic paint on their backs to determine the winner and loser of each contest. From this innovation, it was clear that losers emitted the compound, probably as an unpleasant smell to confuse the winner while the loser flew away. To ensure that the ion at m/z 171 was detecting 2-methyl-1,7-dioxaspiro[5.5]undecane, the cone voltage on the APCI-MS was changed to induce some fragmentation of the molecule. The fragmentation pattern confirmed the identity of the compound (76).

A Self-Service APCI-MS System for Plant Breeding Studies

Tomato Introgression Lines

Following the success with monitoring insect behavior, colleagues in the Plant Sciences Division were keen to find a rapid method to assess the flavor quality of tomatoes. They were aware that tomato metabolism around maturity changes very rapidly and could cause serious variation in their data. Ideally, they wanted to pick a single tomato from their environmentally-controlled glasshouse at a well-defined time post-flowering, so that metabolic variation was minimized. It was envisaged that several hundred tomatoes would be harvested over the season and on a 7-day a week, 12-hour day schedule. GC-MS analysis was clearly too time-consuming and also involved considerable sample preparation. To meet the needs of our colleagues, we designed a simple maceration device (coffee blender; Figure 6A) that was connected to the direct APCI-MS system to make it possible to analyze volatile compounds from a single tomato (28). The APCI-MS was set up with *m/z* values for the key volatiles of interest and had an analysis time of 5 min per tomato, with the system operated by Plant Science researchers in our laboratory as a "walk up" or "self-service" facility.

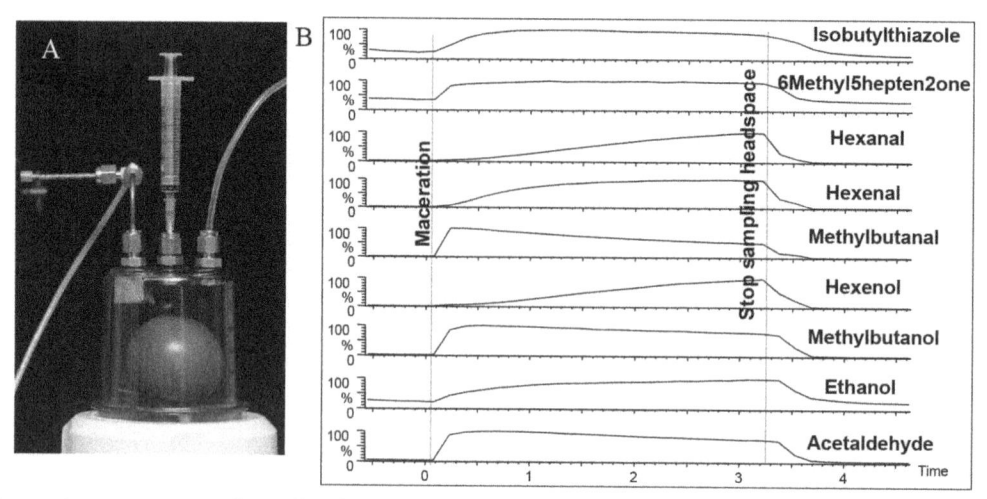

Figure 6. A: Maceration device for releasing tomato aromas for APCI-MS analysis. B: Traces from direct APCI-MS of tomatoes before and after maceration. Note some compounds are present before maceration and increase over the 5 min analysis time, others are only produced in detectable amounts after maceration.

A typical volatile trace is shown in Figure 6B. The characteristic aroma of tomatoes is derived from some compounds formed in the fruit during ripening and other compounds that are formed only when the fruit is disrupted and decompartmentalization of enzymes and substrates occurs to produce fresh-smelling compounds like hexenal, hexanal and 6-methyl-5-hepten-2-one. This evolution of compounds can be seen in Figure 6B. From the rate of volatile production for these compounds, an estimate of enzyme activity in each tomato could be derived. The outcome of the self-service analysis was that over 600 individual tomato fruits were analyzed from three introgression lines. Comparison of the relative amounts of volatiles produced during maceration was used to define the phenotype of the different tomato lines developed.

The rapid tomato aroma analysis system was also adapted to carry out more fundamental studies on the enzyme-derived aroma compounds. The studies were in response to questions from plant breeders who wished to manipulate the aroma profile in tomatoes and wanted to know what were the limiting factors in enzymatic tomato aroma production, for example, was it the enzyme activity or the availability of substrate for the enzymes. By adding small quantities of substrates like linoleic acid, linolenic acid and terpenes to the intact tomato before blending, the effect of additional substrate could be determined (77).

Conclusion

Although APCI-MS was developed with a key application in mind, the simplicity of the technique and the imagination of many collaborators has led to a host of other applications. Despite the fact that APCI-MS is not always as sensitive as other direct MS methods, its strengths are its flexibility and the ease of adapting it to cope with such a variety of applications with some simple "plumbing" connections. Unfortunately, commercialization of the technique was limited due to the takeover of Micromass by Waters and their understandable focus on high-end mass spectrometers rather than interfaces for existing models. However, the Nottingham team feel that their extensive published work on APCI applications (over 100 publications) showed that direct APCI-MS was, and is still, a valuable tool for commercial and academic research laboratories.

Acknowledgments

Dr Tony Blake (Firmenich SA) was instrumental in supporting the development of APCI-MS at Nottingham in the 1990s. His infectious enthusiasm led to a joint project with Firmenich, Unilever, Mars and Defra that established the key features of the technique. Dr Rob Linforth was responsible for the design of the APCI-MS interface (also known as MS Nose), and was also responsible for training generations of students in its use and inspiring exceptional creativity in the lab.

References

1. Taylor, A. J. Release and transport of flavours in vivo: physico chemical, physiological and perceptual considerations. *Comp. Rev. Food Safety Food Sci.* **2002**, *1*, 45–57.
2. Brauss, M. S.; Linforth, R. S. T.; Cayeux, I.; Harvey, B.; Taylor, A. J. Altering the fat content affects flavour release in a model yogurt system. *J. Agric. Food Chem.* **1999**, *47*, 2055–2059.
3. Kant, A.; Linforth, R. S.; Hort, J.; Taylor, A. J. Effect of beta-cyclodextrin on aroma release and flavor perception. *J. Agric. Food Chem.* **2004**, *52*, 2028–2035.
4. Bayarri, S.; Taylor, A. J.; Hort, J. The Role of Fat in Flavor Perception: Effect of Partition and Viscosity in Model Emulsions. *J. Agric. Food Chem.* **2006**, *54*, 8862–8868.
5. Carey, M.; Linforth, R. S. T.; Taylor, A. J. In *Flavour Research at the dawn of the Twenty-first Century*; LeQuere, J.-L., Etievant, P. X., Eds.; Lavoisier: London, 2003; pp 212–215.
6. Doyen, K.; Carey, M.; Linforth, R. S. T.; Marin, M.; Taylor, A. J. Volatile release from an emulsion: Headspace and in-mouth studies. *J. Agric. Food Chem.* **2001**, *49*, 804–810.
7. Linforth, R. S. T.; Taylor, A. J. In *Food Lipids: chemistry, flavor and texture*; Shahidi, F., Weenen, H., Eds.; American Chemical Society: Washington, DC, 2006; Vol. 920, pp 159–170.
8. Cook, D. J.; Linforth, R. S. T.; Taylor, A. J. Effects of hydrocolloid thickeners on the perception of savory flavors. *J. Agric. Food Chem.* **2003**, *51*, 3067–3072.
9. Hatakeyama, J.; Davidson, J. M.; Kant, A.; Koizumi, T.; Hayakawa, F.; Taylor, A. J. Optimizing aroma quality in curry sauce products using in vivo aroma release measurements. *Food Chem.* **2014**, *157*, 229–239.
10. González-Tomás, L.; Bayarri, S.; Taylor, A. J.; Costell, E. Flavour release and perception from model dairy custards. *Food Res. Int.* **2007**, *40*, 520–528.
11. González-Tomás, L.; Bayarri, S.; Taylor, A. J.; Costell, E. Rheology, flavour release and perception of low-fat dairy desserts. *Int. Dairy J.* **2008**, *18*, 858–866.
12. Brauss, M. S.; Cayeux, I.; Harvey, B.; Linforth, R. S. T.; Taylor, A. J. The fat content of yoghurts determines the release of flavour volatiles during eating. *Proceedings of the Nutrition Society* **1998**, *57*, 135A.
13. Pionnier, E.; Nicklaus, S.; Chabanet, C.; Mioche, L.; Taylor, A. J.; Le Quere, J. L.; Salles, C. Flavor perception of a model cheese: relationships with oral and physico-chemical parameters. *Food. Qual. Prefer.* **2004**, *15*, 843–852.

14. Salles, C.; Hollowood, T. A.; Linforth, R. S. T.; Taylor, A. J. In *Flavour Research at the dawn of the Twenty-first Century*; LeQuere, J.-L., Etievant, P. X., Eds.; Lavoisier: London, 2003; pp 170–175.
15. Brauss, M. S.; Balders, B.; Linforth, R. S. T.; Avison, S.; Taylor, A. J. Fat content, baking time, hydration and temperature affect flavor release from biscuits in model and real systems. *Flav. Frag J.* **2000**, *14*, 351–357.
16. Brauss, M. S.; Linforth, R. S. T.; Taylor, A. J. In *Frontiers of Flavour Science*; Schieberle, P., Engel, K.-H., Eds.; Deutsche Forschungsanstalt fuer Lebensmittelchemie: Garching, 2000; pp 279–281.
17. Burseg, K. M. M.; Linforth, R. S. T.; Hort, J.; Taylor, A. J. Flavor Perception in Biscuits; Correlating Sensory Properties with Composition, Aroma Release, and Texture. *Chemosens. Percept.* **2009**, *2*, 70–78.
18. Blee, N.; Linforth, R.; Yang, N.; Brown, K.; Taylor, A. Variation in aroma release between panellists consuming different types of confectionary. *Flav. Frag. J.* **2011**, *26*, 186–191.
19. Blee, N.; Linforth, R. S. T.; Yang, N.; Brown, K.; Taylor, A. J. Effect of sample type and the air/water partition coefficient, on variation in in-vivo flavour release. *Flav. Frag. J.* **2011**, *26*, 186–191.
20. Blissett, A.; Hort, J.; Taylor, A. J. Influence of chewing and swallowing behavior on volatile release in two confectionery systems. *J. Texture Studies* **2006**, *37*, 476–496.
21. Harvey, B. A.; Davidson, J. M.; Linforth, R. S. T.; Taylor, A. J. In *Frontiers of Flavour Science*; Schieberle, P., Engel, K.-H., Eds.; Deutsche Forschungsanstalt fuer Lebensmittelchemie: Garching, 2000; pp 271–274.
22. Jones, L. L.; Chu, K.; Wollmann, N.; Taylor, A. J.; Skelton, S. In *Current Topics in Flavor Chemistry & Biology. Proceedings of the 10th Wartburg Symposium*; Hoffman, T., Krautwurst, D., Schieberle, P., Eds.; Deutsche Forschungsanstalt fuer Lebensmittelchemie: Freising, 2014; pp 461–464.
23. Linforth, R. S. T.; Baek, I.; Taylor, A. J. Simultaneous instrumental and sensory analysis of volatile release from gelatine and pectin/gelatine gels. *Food Chem.* **1999**, *65*, 77–83.
24. Linforth, R. S. T.; Pearson, K. S.-K.; Taylor, A. J. In-vivo flavor release from gelatin-sucrose gels containing droplets of flavor compounds. *J. Agric. Food Chem.* **2007**, *55*, 7859–7863.
25. Pfeiffer, J.; Hort, J.; Hollowood, T. A.; Taylor, A. J. Taste-Aroma Interactions in a Ternary System: a Model of Fruitiness Perception in Sucrose/Acid Solutions. *Perception & Psychophysics* **2006**, *68*, 216–227.
26. Taylor, A. J.; Besnard, S.; Puaud, M.; Linforth, R. S. T. In vivo measurement of flavour release from mixed phase gels. *Biomolecular Engineering* **2001**, *17*, 143–150.
27. Yang, N.; Linforth, R. S. T.; Walsh, S.; Brown, B.; Hort, J.; Taylor, A. J. Feasibility of reformulating flavours between food products using in vivo aroma comparisons. *Flav. Frag, J.* **2011**, *26*, 107–115.
28. Boukobza, F.; Dunphy, P.; Taylor, A. J. Measurement of lipid oxidation-derived volatiles in fresh tomatoes. *Postharvest Biology and Technology* **2001**, *23*, 117–131.
29. Boukobza, F.; Taylor, A. J. In *Freshness and shelf-life of foods*; Weenen, H., Cadwallader, K., Eds.; American Chemical Society: Washington, DC, 2002; Vol. 836, pp 132–143.

30. Friel, E. N.; Wang, M.; Taylor, A. J.; Macrae, E. Real-time release of flavor components of gold fleshed kiwifruit (Actinidia chinensis). *J. Agric. Food Chem.* **2007**, *55*, 6664–6673.
31. Hodgson, M. D.; Langridge, J. P.; Linforth, R. S. T.; Taylor, A. J. Aroma Release and Delivery Following the Consumption of Beverages. *J. Agric. Food Chem.* **2005**, *53*, 1700–1706.
32. Miettinen, S. M.; Hyvonen, L.; Linforth, R. S. T.; Taylor, A. J.; Tuorila, H. Temporal aroma delivery from milk systems containing 0-5% added fat, observed by free choice profiling, time intensity, and atmospheric pressure chemical ionization-mass spectrometry techniques. *J. Agric. Food Chem.* **2004**, *52*, 8111–8118.
33. Rabe, S.; Linforth, R. S. T.; Krings, U.; Taylor, A. J.; Berger, R. G. Flavour Release from Liquids: A Comparison of In Vivo APCI-MS, In Mouth Headspace Trapping, and In Vitro Mouth Model Data. *Chem. Senses* **2004**, *29*, 163–173.
34. Taylor, A. J.; Tsachaki, M.; Lopez, R.; Morris, C.; Ferreira, V.; Wolf, B. In *Flavors in noncarbonated beverages*; Da Costa, N., Cannon, R. J., Eds.; ACS Symposium Series: Washington, DC, 2010; Vol. 1036, pp 161–175.
35. Tsachaki, M.; Aznar, M.; Linforth, R. S. T.; Taylor, A. J. In *Flavour Science: Recent advances and trends*; Bredie, W. L. P., Petersen, M. A., Eds.; Elsevier: Amsterdam, 2006; Vol. 43, pp 441–444.
36. Tsachaki, M.; Gady, A.-L.; Kalopesas, M.; Athes, V.; Linforth, R. S. T.; Marin, M.; Taylor, A. J. Effect of ethanol, temperature and gas flow rate on volatile release from aqueous solutions under dynamic headspace dilution conditions. *J. Agric. Food Chem.* **2008**, *56*, 5308–5315.
37. Tsachaki, M.; Linforth, R. S. T.; Taylor, A. J. Dynamic Headspace Analysis Of The Release Of Volatile Organic Compounds From Ethanolic Systems By Direct APCI-MS. *J. Agric. Food Chem.* **2005**, *53*, 8328–8333.
38. Tsachaki, M.; Linforth, R. S. T.; Taylor, A. J. Aroma Release from Wines under Dynamic Conditions. *J. Agric. Food Chem.* **2009**, *57*, 6976–6981.
39. Wright, J.; Wulfert, F.; Hort, J.; Taylor, A. J. Effect of Preparation Conditions on Release of Selected Volatiles in Tea Headspace. *J. Agric. Food Chem.* **2007**, *55*, 1445–1453.
40. Wright, K. M.; Hills, B. P.; Hollowood, T. A.; Linforth, R. S. T.; Taylor, A. J. Persistence effect in flavour release from liquids in the mouth. *Int. J. Food Sci. Tech.* **2003**, *38*, 343–350.
41. Aznar, M.; Tsachaki, M.; Linforth, R. S. T.; Ferreira, V.; Taylor, A. J. Headspace analysis of volatile organic compounds from ethanolic systems by direct APCI-MS. *Int. J. Mass Spec.* **2004**, *239*, 17–25.
42. Davidson, J. M.; Hollowood, T. A.; Linforth, R. S. T.; Taylor, A. J. The effect of sucrose on the perceived flavour intensity of chewing gum. *J. Agric. Food Chem.* **1999**, *47*, 4336–4340.
43. Hort, J.; Hollowood, T. A. Controlled Continuous Flow Delivery System for Investigating Taste-Aroma Interactions. *J. Agric. Food Chem.* **2004**, *52*, 4834–4843.
44. Cook, D. J.; Davidson, J. M.; Linforth, R. S. T.; Taylor, A. J. In *Handbook of Flavor Characterization. Sensory Analysis, Chemistry and Physiology*; Deibler, K. D., Delwiche, J., Eds.; Marcel Dekker: 2003; pp 135–150.
45. Morris, C.; Labarre, C.; Koliandris, A.-L.; Hewson, L.; Wolf, B.; Taylor, A. J.; Hort, J. Effect of pulsed delivery and bouillon base on saltiness and bitterness perceptions of salt delivery profiles partially substituted with KCl. *Food. Qual. Prefer.* **2010**, *21*, 489–494.

46. Pfeiffer, J. C.; Hollowood, T. A.; Hort, J.; Taylor, A. J. Temporal synchrony and integration of sub-threshold taste and smell signals. *Chem. Senses* **2005**, *30*, 539–45.

47. Cook, D. J.; Hollowood, T. A.; Linforth, R. S. T.; Taylor, A. J. Perception of taste intensity in solutions of random coil polysaccharides above and below C*. *Food Qual. Pref.* **2002**, *13*, 473–480.

48. Cook, D. J.; Hollowood, T. A.; Linforth, R. S. T.; Taylor, A. J. Oral shear stress predicts flavour perception in viscous solutions. *Chem. Senses* **2003**, *28*, 11–23.

49. Cook, D. J.; Hollowood, T. A.; Linforth, R. S. T.; Taylor, A. J. Correlating instrumental measurements of texture and flavour release with human perception. *Int. J. Food Sci. Tech.* **2005**, *40*, 631–641.

50. Brossard, C. D.; Lethuaut, L.; Boelrijk, A. E. M.; Mariette, F.; Genot, C. Sweetness and aroma perceptions in model dairy desserts: an overview. *Flav. Frag. J.* **2006**, *21*, 48–52.

51. Salles, C.; Engel, E.; Nicklaus, S.; Taylor, A. J.; Le Quéré, J. L. In *Process and reaction flavours*; Weerasinghe, D. K., Sucan, M. K., Eds.; American Chemical Society: Washington, DC, 2005; Vol. 905, pp 192−207.

52. Mestres, M.; Moran, N.; Jordan, A.; Buettner, A. Aroma release and retronasal perception during and after consumption of flavored whey protein gels with different textures. 1. in vivo release analysis. *J. Agric. Food Chem.* **2005**, *53*, 403–409.

53. Ferry, A. L.; Hort, J.; Mitchell, J. R.; Lagarrigue, S.; Pamies, B. Effect of amylase activity on starch paste viscosity and its implications for flavor perception. *J. Texture Studies* **2004**, *35*, 511–524.

54. Malone, M. E.; Appelqvist, I. A. M.; Norton, I. T. Oral behaviour of food hydrocolloids and emulsions. Part 2. Taste and aroma release. *Food Hydrocolloids* **2003**, *17*, 775–784.

55. Palsgard, E.; Dijksterhuis, G. The sensory perception of flavor release as a function of texture and time: A time intensity study using flavored gels. *J. Sens. Stud.* **2000**, *15*, 347–359.

56. Le Quere, J. L.; Gierczynski, I.; Semon, E. An atmospheric pressure chemical ionization-ion-trap mass spectrometer for the on-line analysis of volatile compounds in foods: a tool for linking aroma release to aroma perception. *J. Mass Spec.* **2014**, *49*, 918–928.

57. Beauchamp, J.; Zardin, E. Odorant Detection by On-line Chemical Ionization Mass Spectrometry. In *Handbook of Odor*; Buettner, A., Ed.; 2017; pp 55−408.

58. Biasioli, F. In *PTR-MS in food science and technology: a review*; Hansel, A., Mark, T. D., Eds.; Innsbruck University Press: Obergurgl, 2007; pp 111−115.

59. Semon, E.; Arvisenet, G.; Guichard, E.; Le Quere, J. L. Modified proton transfer reaction mass spectrometry (PTR-MS) operating conditions for in vitro and in vivo analysis of wine aroma. *J. Mass Spec.* **2018**, *53*, 65–77.

60. Langford, V. S.; Padayachee, D.; McEwan, M. J.; Barringer, S. A. Comprehensive odorant analysis for on-line applications using selected ion flow tube mass spectrometry (SIFT-MS). *Flav. Frag. J.* **2019**, *34*, 393–410.

61. Zhang, Y.; Barringer, S. Effect of hydrocolloids, sugar, and citric acid on strawberry volatiles in a gummy candy. *Food Process. Preserv.* **2018**, *42*, e13327.

62. Linforth, R. S. T.; Cabannes, M.; Yang, N.; Taylor, A. J.; Hewson, L. Effect of fat content on flavor delivery during consumption: an in-vivo model. *J. Agric. Food Chem.* **2010**, *58*, 6905–6911.

63. Linforth, R. S. T.; Friel, E. N.; Taylor, A. J. In *Flavor release: Linking experiments, theory and reality*; Roberts, D. D., Taylor, A. J., Eds.; American Chemical Society: Washington, DC, 2000; Vol. 763, pp 166–178.
64. Hodgson, M.; Linforth, R. S. T.; Taylor, A. J. Simultaneous Real Time Measurements of Mastication, Swallowing, Nasal Airflow and Aroma Release. *J. Agric. Food Chem.* **2003**, *51*, 5052–5057.
65. Shojaei, Z. A.; Linforth, R. S. T.; Hort, J.; Hollowood, T. A.; Taylor, A. J. Measurement and manipulation of aroma delivery allows control of perceived fruit flavour in low and regular fat milks. *Int. J. Food Sci. Tech.* **2006**, *41*, 1192–1196.
66. Marin, M.; Baek, I.; Taylor, A. J. Flavour release from aqueous solutions under dynamic headspace dilution conditions. *J. Agric. Food Chem.* **1999**, *47*, 4750–4755.
67. Friel, E. N.; Linforth, R. S. T.; Taylor, A. J. An empirical model to predict the headspace concentration of volatile compounds above solutions containing sucrose. *Food Chem.* **2000**, *71*, 309–317.
68. Choi, S.; Parameswaran, S.; Choi, J.-H. Understanding alcohol aggregates and the water hydrogen bond network towards miscibility in alcohol solutions: graph theoretical analysis. *Physical Chemistry Chemical Physics* **2020**, *22*, 17181–17195.
69. Spedding, P. L.; Grimshaw, J.; O'Hare, K. D. Abnormal Evaporation Rate of Ethanol from Low Concentration Aqueous-Solutions. *Langmuir* **1993**, *9*, 1408–1413.
70. Stewart, E.; Shields, R. L.; Taylor, R. S. Molecular dynamics simulations of the liquid/vapor interface of aqueous ethanol solutions as a function of concentration. *Journal of Physical Chemistry B* **2003**, *107*, 2333–2343.
71. Hodge, J. E. Dehydrated Foods, Chemistry of Browning Reactions in Model Systems. *J. Agric. Food Chem.* **1953**, *1*, 928–943.
72. Channell, G. A.; Taylor, A. J. In *Process and Reaction Flavors*; Weerasinghe, D. K., Sucan, M. K., Eds.; American Chemical Society: Washington, DC, 2005; Vol. 905, pp 181–191.
73. Channell, G. A.; Wulfert, F.; Taylor, A. J. Identification and monitoring of intermediates and products in the acrylamide pathway using on-line analysis. *J. Agric. Food Chem.* **2008**, *56*, 6097–6104.
74. Taylor, A. J.; Sivasundaram, L.; Moreau, L.; Channell, G. A.; Hill, S. E. In *Controlling Maillard Pathways To Generate Flavors*; Mottram, D. S., Taylor, A. J., Eds.; American Chemical Society: Washington, DC, 2010; Vol. 1042, pp 129–142.
75. Goubault, M.; Batchelor, T. P.; Linforth, R. S. T.; Taylor, A. J.; Hardy, I. C. W. Volatile emission by contest losers revealed by real-time chemical analysis. *Proceedings of the Royal Society of London Series B-Biological Sciences* **2006**, *273*, 2853–2859.
76. Goubault, M.; Batchelor, T. P.; Romani, R.; Linforth, R. S. T.; Fritzsche, M.; Francke, W.; Hardy, I. C. W. Volatile chemical release by bethylid wasps: identity, phylogeny, anatomy and behaviour. *Biological Journal of the Linnean Society* **2008**, *94*, 837–852.
77. Boukobza, F.; Taylor, A. J. In *Freshness and shelf-life of foods*; Weenen, H., Cadwallader, K., Eds.; American Chemical Society: Washington, DC, 2002; Vol. 836, pp 124–131.

Chapter 3

Pushing the Boundaries of Dynamic Flavor Analysis with PTR-MS

Jonathan D. Beauchamp[*]

Department of Sensory Analytics and Technologies, Fraunhofer Institute for Process Engineering and Packaging IVV, Giggenhauser Str. 35, 85354 Freising, Germany
[*]Email: jonathan.beauchamp@ivv.fraunhofer.de

Proton transfer reaction-mass spectrometry (PTR-MS) is an analytical technique that detects volatile organic compounds (VOCs), amongst them flavor/aroma compounds, in the gas-phase. The instrument is constructed and configured in a manner that allows rapid and continuous compound detection over a broad dynamic range. These features have positioned PTR-MS as an ideal measurement tool in food science to characterize dynamic processes, such as flavor release or food spoilage. Since it emerged on the analytical chemistry landscape more than a quarter century ago, the capabilities of PTR-MS have been exploited in numerous food/flavor research applications to complement the conventional approach of gas chromatography-mass spectrometry (GC-MS). PTR-MS boasts a comprehensive back catalog of studies on all manner of foods and beverages, ranging from coffee, to fruits, meat, dairy and oils. Assessments span volatile (flavor) composition screening, dynamic flavor release, spoilage, fermentation or maturation process monitoring, high-throughput profiling, model food systems, and *in vivo* release, amongst others. This chapter presents an overview of the development and uses of PTR-MS in food/flavor research, from its historic first steps to present day advanced applications.

A Long Time Ago in a Valley Far, Far Away...

Proton transfer reaction-mass spectrometry (PTR-MS) emerged from the basement laboratories of the Institute of Ion Physics at the University of Innsbruck, Austria, in the mid-1990s (*1*). The technique derived from bulkier instrumentation with a large laboratory footprint, the selected ion flow drift tube-mass spectrometer (SIFDT-MS), which itself arose as a culmination of developments dating back to the 1960s that commenced with the observations of the flowing afterglow phenomenon (*2*) and the discovery of chemical ionization (*3*). The combination of a flowing afterglow system with a drift tube (*4*) and a Venturi inlet (*5*) led to the invention of SIFDT-MS – a close relative to selected ion flow tube-mass spectrometry (SIFT-MS); see Chapter 4 – that

© 2021 American Chemical Society

was utilized extensively to study ion-molecule interactions, amongst them proton transfer reactions (6). These studies – and similar work using SIFT-MS to explore ion-neutral reactions in the terrestrial atmosphere and interstellar clouds (7) – generated a wealth of data on reaction rate coefficients between selected charged reagent ions and neutral analyte molecules. With this empirical data at hand, researchers soon realized that this technology could be used in the opposite manner, i.e., to introduce a gas sample of unknown chemical composition into the instrument and exploit the amassed coefficient values to estimate ion densities of individual analytes in the reaction chamber, thereby enabling the gas-phase quantities of volatile organic compounds (VOCs) in the original gas sample to be derived (8, 9). This constituted the dawn of a new era in analytical chemistry: real-time mass spectrometry.

Returning to the River Inn valley in the Tyrol region of Austria in the mid-1990s, the team of physicists at the University of Innsbruck, led by Werner Lindinger, adapted the unwieldy SIFDT-MS apparatus via a series of modifications in view of developing this novel application of on-line mass spectrometric gas analysis. A pivotal development was the predilection of using protonated water, i.e., H_3O^+, as the sole reagent ion and to tailor this new system accordingly. Ion-molecule reactions using H_3O^+ – or more specifically, $H_2O \cdot H^+$ – proceed via the transfer of a proton from this reagent ion to the neutral analyte, yielding a protonated product ion. The use of H_3O^+ as the proton donor in this reaction has three distinct benefits: first, the energy of the proton transfer reaction is low (generally <2 eV) and thus only slightly exothermic, meaning that it proceeds mostly non-dissociatively, i.e., without breakup of the neutral target, in other words, with little or no fragmentation; second, the ubiquity of water (vapor) in gas samples presents a challenge in many analytical applications as it disturbs sampling and is difficult to remove without compromising the sample, but by using H_3O^+ as a reagent ion water vapor was no longer considered an interference (nor represented an impurity in the ion source); third and finally, the proton affinity (PA) of water (see Chapter 1 and Figure 1 therein), conveniently, is lower than that of most VOCs but higher than the PAs of the common constituents of air (nitrogen, oxygen, argon, carbon dioxide), therefore protonated water does not react with these components and, consequently, air can be used in the flow-drift tube reactor. The latter means that air samples can be analyzed directly, without dilution, thereby achieving very low detection limits; latest commercial instruments perform down to sub-ppt detection under ideal conditions (10).

The selection of operating only with one reagent ion (H_3O^+) allowed a further decisive modification by the Innsbruck team: the replacement of the front-end mass spectrometer – used to pre-selected reagent ions for injection in the reaction chamber in SIFDT-MS – with a cylindrical hollow cathode discharge ion source that generated H_3O^+ ions of high purity (≥99.5 %) from a constant and controlled flow of water vapor through the source (11). This alteration, as well as a smaller geometry of the flow-drift tube, enabled an overhaul of the vacuum pumping system, which allowed this new system to be greatly reduced in size compared to the SIFDT-MS instrument. A new instrument was born and was given the name proton transfer reaction-mass spectrometer (PTR-MS) (1). For the sake of completeness, it is noteworthy that subsequent developments have led to the optional use of NO^+ or O_2^+ as reagent ions in PTR-MS, as is similarly employed in SIFT-MS (see Chapter 1), but the use of this feature, termed switchable (or selectable) reagent ion (SRI) (12), has found limited implementation in food/flavor applications to date and is thus not discussed further in this chapter.

Heralding a New Era in Food/Flavor Analysis

The newly developed PTR-MS instrument presented abundant opportunities in VOC analysis. Indeed, the first pioneering studies, carried out by Werner Lindinger, Armin Hansel, Alfons Jordan and co-workers, explored a variety of applications relevant to different fields, from ambient air quality monitoring, to exhaled breath-biomarker analysis, to aroma release assessments, amongst others (*1, 13, 14*). Focusing on its implementation in food analysis, the first measurements by PTR-MS in this area were proof-of-principle studies that demonstrated the utility of this new tool for food-related applications. These included a study of the temporal release of methanol, ethanol, acetaldehyde, acetic acid, propanol and acetone from different berries (strawberries, white currant, raspberries and blackberries), acetic acid, 2,3-butanedione/2-methylbutanal, furfural, acetaldehyde, acetone/propanal, methylethylketone/methylpropanal, and ethylformeate/methylacetate from freshly ground coffee, and methanethiol, dimethylsulfide, acetaldehyde, dimethylamine and methanol from meat during spoilage (*14*).

These studies provided the first indications of the unique capabilities of PTR-MS to characterize dynamic processes in food systems, heralding a new era in food science. In particular, the on-line detection afforded by PTR-MS technology offered major benefits compared to conventional aroma analysis, e.g., by gas chromatography-mass spectrometry (GC-MS). First, the ability to sample gas on-line obviated the need for laborious sample workup procedures before analysis. Consequently, samples could be measured directly and non-destructively by means of headspace analysis, which reduced potential losses or changes in sample composition prior to analysis and offered the possibility of repeated analysis per sample. As a further consequence, the short measurement intervals permitted by on-line analysis allowed the quantitative and qualitative changes of the VOC composition of a sample to be examined over time, which was ideal for assessing rapid aroma release kinetics or gradual food spoilage processes. It should be noted that these early studies were concurrent with the development and exploration of atmospheric pressure chemical ionization-mass spectrometry (APCI-MS) (see Chapters 1 and 2) for food science applications, with open academic discussion between the two camps that was of mutual benefit to both sides. It is appropriate to mention here that the original PTR-MS instruments were equipped with a quadrupole mass filter, thus at least the first decade of PTR-MS studies were limited to the use of this system, which offered only nominal mass resolution; the replacement of this mass filter with a time-of-flight (TOF) mass spectrometer is discussed later. Nowadays, the two systems are commonly distinguished by referring to them as PTR-QMS and PTR-TOF-MS, or generically as PTR-MS.

Following these early pilot studies, the use of PTR-MS for food/flavor analysis gained momentum when it was adopted by a handful of food scientists who worked closely with the PTR-MS inventors to explore its potential for a variety of applications. Notably, a collaboration with the Agri-Food Quality Department at the Istituto Agrario di San Michele all'Adige (now Fondazione Edmund Mach; FEM) in Italy yielded a rich portfolio of application notes and research articles that demonstrated the unique possibilities offered by this new technology. Spanning the late 1990s to mid-2000s, the Italian group, centered around Franco Biasioli, Eugenio Aprea and Flavia Gasperi, together with their collaborators in Austria, published a series of diverse studies that ranged from monitoring post-harvest aging of berry fruits (*15*), exploring *in vivo* flavor release of flavored model food systems in relation to oral processing (*16*), or comparing flavor profiles of different cheeses acquired by PTR-MS and GC-olfactometry (GC-O) analyses (*17*). Of particular note is the groundbreaking work undertaken by this group relating to the concept of mass spectral fingerprinting and chemometrics, whereby multivariate statistical models were applied to the PTR-MS data to tease out nuances between samples. The novelty of this approach was that the identification of the individual compounds within the PTR-MS mass spectra of sample headspace was unnecessary, since the mathematical comparison of the data was based on quantitative shifts in the signal patterns across different mass spectra, rather than focusing on individual compounds *per se*.

PTR-MS was predestined to flourish in this emerging application due to its non-destructive and high-throughput capabilities. Correspondingly, the Italian team explored the limits of this approach, with studies assessing food processing, such as the heat and/or pressure treatment of orange juice (*18*), a comparison of different strawberry varieties (*19, 20*) or apple genotypes (*21*), assessments of the degree of oxidation alteration of olive oil (*22*), and a correlative evaluation of the sensory profiles and mass spectral fingerprints of Trentingrana cheese (*23, 24*), amongst other studies. The more recent implementation by this group of an autosampling system for high-throughput PTR-TOF-MS analysis, especially in exploring flavor development during fermentation, is worthy of special mention and is discussed later.

In addition to this early collaborative work by the Austrian and Italian teams, the new PTR-MS technology found interest at the Nestlé Research Center in Lausanne, Switzerland, with a focal application in coffee. Indeed the adoption of PTR-MS by Nestlé's research division provided essential support in establishing this tool as an innovative and robust commercial analytical system for food/flavor applications. The collaboration between Werner Lindinger's team in Innsbruck and Chahan Yeretzian, Philippe Pollien and co-workers in Lausanne led to the publication of several key studies. These included a seminal work reporting on a comprehensive examination of volatile aroma compounds in coffee, from the headspace of green beans, to the aroma generation during coffee bean roasting (single bean or batch roasting), as well as the emissions during coffee brewing (*25*). A companion study on the headspace analysis of coffee by PTR-MS contained an extensive comparison of known aroma compounds in coffee with the signals detected by PTR-MS, with a corresponding examination of the fragmentation patterns of target analytes, as well as an exploration of the depletion of selected VOCs in the headspace of brewed coffee in relation to the partitioning from the liquid to gas phase (*26*). The latter phenomenon was further explicitly investigated in a separate study, using a similar approach, in which the innovative method was refined to allow partition coefficients (or Henry's law constants) to be derived empirically (*27, 28*).

Other notable studies from this collaboration included analyses that compared headspace profiles with *in vivo* flavor release, the latter being termed 'nosespace' analysis due to sampling by PTR-MS at the nostril of the panelist (*29, 30*), and on-line monitoring of the Maillard reaction, for example, with a focus on the generation of acrylamide (*31*). In recent years, Yeretzian (presently at the Zurich University of Applied Sciences in Switzerland) and colleagues have exploited the analytical strengths of PTR-MS, particularly PTR-TOF-MS, to the fullest extent in their pursuits to decipher coffee flavor, from bean to brew, including roasting (*32*). Their innovative work on characterizing the kinetics of coffee aroma compound generation during extraction, as analyzed directly at the coffee machine via a bespoke sampling interface, and with multivariate data processing using principal component analysis (PCA) and hierarchical cluster analysis (HCA), provides new insights into coffee flavor development in relation to serving size and extraction parameters, such as temperature and pressure (*33, 34*). Other work on coffee by this group has targeted roasted coffee oil and explored the impact of ultrasound-assisted emulsification and microencapsulation on dynamic aroma release (*35, 36*). Finally, in an unrelated yet equally insightful study, Yeretzian and co-workers undertook *in vivo* measurements on aroma persistent for a variety of compounds in view of exploring the 'after-taste' (or 'after-odor') phenomenon (*37*), as is also treated in Chapter 5. The study examined compounds that exhibited a range of physicochemical properties, with the *in vivo* analyses indicating that polarity and vapor pressure were the driving factors for interactions of the aroma compounds with the airway mucosa, thus having the greatest influence on after-taste.

Towards the mid-2000s, PTR-MS started being adopted by other food science groups that were pursuing divergent applications. Saskia van Ruth, initially at University College Cork in Ireland, then at Wageningen University and Research in the Netherlands, conducted broad-ranging investigations using the PTR-MS instrumentation, including studies characterizing the performance of the instrument for the detection of aroma compounds (38) and a comparison of its performance to GC (39), the release and intranasal distribution of aroma compounds from model systems (40–43), and the first coupling of a PTR-MS instrument to a model mouth for the systematic assessment of simulated food oral processing (44). One specific application pioneered by van Ruth and team was the use of mass spectral fingerprinting by PTR-MS, in combination with multivariate statistics, to discriminate food samples in relation to classification, geographic origin and/or authenticity, including butter (45, 46), cheese (47) and oils (48, 49).

Further early work on PTR-MS in food science applications came from Jean-Luc Le Quéré, Isabelle Souchon, Elizabeth Guichard and team at the Centre des Sciences du Goût et de l'Alimentation at the French National Research Institute for Agriculture, Food and Environment (INRAE) in Dijon, France, who were forerunners in developing and comparing sensory methods with *in vivo* nosespace analysis by PTR-MS (50, 51) (and who, incidentally, contributed significantly to the APCI-MS literature, too (52, 53)), and Andrea Buettner at the Technical University of Munich, Germany (now at Fraunhofer IVV, Freising, Germany), who examined the relationship between retronasal flavor perception and physiological processes (54–57), amongst other studies.

More recently, the activities of Buettner and colleagues at Fraunhofer IVV, including the author of this chapter, have utilized PTR-MS in their pursuits of diverse research goals. Specifically, *in vivo* (intranasal) analyses were applied to ascertain the transfer of aroma-active compounds to the nasal epithelium in relation to sniff behavior and odor perception (58) and the performances of odor delivery tools for related studies were assessed by direct analysis using PTR-MS (59, 60). Further, the group have carried out extensive work on flavor release in relation to food matrix ingredients, such as sugar and sweetners (61–63), and on food spoilage mechanisms and the potential to monitor such processes via signature volatile metabolites. In relation to the latter, and working in collaboration with Patrick Silcock, Phil Bremer and colleagues at the University of Otago (Dunedin, New Zealand), studies have included investigating the dynamic generation and release of VOCs during photooxidation of milk (64, 65), microbial spoilage of milk (66), pork (67, 68), lamb (69) and chicken (70), and, unrelated to spoilage, measurements with a model mouth system to explore simulated aroma release under defined conditions (71, 72, 73).

The dissemination of emerging food/flavor research applications of PTR-MS in its early years – and concurrently, APCI-MS (see Chapter 2) – in the scientific literature and at academic conferences demonstrated its potential to examine food-related processes, such as flavor release, in a hitherto inaccessible manner, and provided traction for its addition to the analytical instrumentation arsenals of food science laboratories, both in academia and industry. By the early 2000s, the number of researchers employing PTR-MS across many disciplines had reached a critical mass to warrant the inauguration of a user community meeting to present data and exchange ideas. The 1st International Conference on Proton Transfer Reaction-Mass Spectrometry and its Applications was held from 18-23 January 2003 in Igls/Innsbruck, Austria and covered topics on atmospheric chemistry, breath analysis for medical applications, and food science, the latter comprising eight talks and nine posters. This debut conference, celebrating the achievements of PTR-MS in diverse research disciplines, was shadowed by the absence of PTR-MS pioneer Werner Lindinger, who had passed away in a tragic accident two years earlier; the conference was dedicated to his memory. Taking place biennially until 2013, then every three years, the most recent meeting held in 2019 constituted the eighth instalment.

Outside-the-Box PTR-MS Highlights

Numerous food science-related studies employing PTR-MS were undertaken in the 'noughties' (i.e., 2000-2009), foremost in applications that exploited its on-line capabilities, such as non-destructive mass spectral fingerprinting analyses, model mouth systems, and *in vivo* flavor release, as reported above. A full account of these studies is beyond the scope of this chapter, but a handful of studies are recounted here. Returning to the theme of coffee, experiments by the aforementioned team at Nestlé sought to explore the ability of PTR-MS to characterize the sensory traits of espresso coffee through rapid headspace analysis. The investigation was led by Christian Lindinger (son of the late Werner Lindinger), who combined the sensory quantitative descriptive analyses (QDA) of coffee samples by a trained panel with their mass spectral fingerprints acquired by PTR-MS (reduced to 16 selected ion traces) to generate a model that could successfully predict the sensory profiles of the different coffees (74). This study is of particular significance, as it demonstrated the possibility for rapid headspace analysis (< 2 min) to deliver a sensory profile of a coffee that traditionally requires extensive evaluation by a panel of highly-trained sensory assessors. A caveat to this approach is that assessments by the trained panel are required as input data to train the model, thus PTR-MS should be considered as a tool to complement the work of a panel, rather than replace it entirely, such as for use in quality control monitoring.

Another application of PTR-MS that was first explored in this period was its use to examine phenomena taking place in the nasal cavity. In a novel study, which was a collaboration between food scientists at University College Cork in Ireland (the eminent and previously-mentioned Saskia van Ruth) and ear, nose and throat (ENT) specialists at the Smell and Taste Clinic at the University of Dresden Medical School (the distinguished 'Godfather of olfaction' Thomas Hummel and his then-colleague Johannes Frasnelli), the PTR-MS sampling inlet tube was placed inside the nose under endoscopic direction and positioned to sample at one of four positions: the nostril, the middle turbinate, the olfactory cleft, and the nasopharynx. Flavored custards of different viscocity were then administered orally and, thus, the concentrations and latency of response of selected aroma compounds could be explored in relation to the sampling location in the nose (41). This highly innovative study yielded several interesting findings, namely that the duration and degree of *in vivo* release of aromas varied between the compounds and that their distributions in the nasal cavity exhibited different latencies and maxima according to the nasal anatomy, as well as other insights in relation to flavor delivery and food matrix consitency. Further, the investigation not only demonstrated a novel use of the instrument, but also highlighted the benefit of interdisciplinary research to attain findings that are of mutual interest to both disciplines; in the present case, flavor science and odor perception. And, it might be added, these measurements corroborated the robustness of the PTR-MS instrument – and the resilience of the PTR-MS operator – in coping with analyses in an extreme environment; mucus and vacuum systems are never a good mix, as this author can testify from first-hand experience (58).

A final study to highlight here relates to the analysis by PTR-MS of alcoholic beverages. Due to the rich aroma composition of alcoholic drinks, such as wine and spirits, their rapid or dynamic analysis by real-time mass spectrometry is of especial interest to academic and industry researchers alike. The high abundance of ethanol in alcoholic drinks, however, presents a challenge for PTR-MS analysis due to a depletion of the H_3O^+ reagent ions and a dominance of protonated ethanol and associated water or ethanol clusters. Although the liquid sample or its headspace can be diluted to prevent this situation, this results in eliminating a large proportion of the aroma compounds of interest, which are typically present at trace quantities (ppm and below). To overcome this problem, Armin Wisthaler and co-workers at the University of Innsbruck explored the possibility to exploit the primary ion redistribution phenomenon occurring in the PTR-MS drift tube in the presence of high concentrations of ethanol by using the ensuing protonated ethanol clusters as reagent ions for the subsequent protonation of trace level aroma constituents (75). This adds a degree of complexity to the resulting mass spectra, but in their study on different wines a fingerprinting approach was used, together with multivariate statistical treatment, as previously described, to distinguish between the wine samples with great success based on their characteristic mass spectra. The innovative approach presented in this study has been adopted and adapted in subsequent studies, such as *in vivo* assessments of wine consumption, delivering highly compelling data (76) and representing a promising method for future studies on other wines and spirits.

Despite this innovative approach of utilizing protonated ethanol as a proton donor for subsequent proton transfer reactions with volatile trace constituents of the beverage, the successes of alcoholic beverage analyses by PTR-MS using the conventional approach should not go unmentioned. In particular, by diluting the sample headspace to reduce the overall concentration of ethanol in the sample gas, the analysis is simplified and can be performed more rapidly, and the interpretation of the ensuing mass spectra is less complex, thus making this approach more suitable to routine applications. Using this method, in combination with multivariate data processing of selected signals in the mass spectra, Wisthaler and colleagues demonstrated the possibility to distinguish between *Pinot Noir* and *Cabernet Sauvignon* samples (77). (As an anecdotal aside here, a co-author of that study asserted to the present author – rather drily, no pun intended – that he could think of an easier method to distinguish between a red and a white wine; but that is beside the point.) A further, recent development in the rapid analysis of sample headspace containing high concentrations of ethanol is to use argon for sample dilution; by using argon, the degree of dilution required is reduced, e.g., 1:3 Ar (78) compared to 1:40 N_2 (77) due to a suppression of ethanol cluster formation and a reduced sensitivity to absolute differences in ethanol concentration between samples (78, 79). Combining this configuration with a fastGC interface, as discussed later, offers yet further benefits.

PTR-TOF-MS: The Final Frontier?

Unlike in GC-MS, in which compounds are identified based on their retention time through a capillary column and their mass spectrum in comparison to those of reference compounds, the detection of individual VOCs by PTR-MS is based solely on the mass of the analyte ion(s). As an example, when protonated, the buttery smelling ketone 2,3-butanedione (diacetyl), which has an elemental composition of $C_4H_6O_2$ and a molecular weight of 86.0892 g/mol, will elicit a signal at a mass-to-charge value (singly charged) of *m/z* 87 in the quadrupole mass filter-based PTR-MS systems, i.e., PTR-QMS. This presents a challenge when additional compounds are present in a sample. Take the compound pentanal, for example, which has an elemental composition of $C_5H_{10}O$ and a molecular weight of 86.1323 g/mol: upon protonation, this fruity smelling aldehyde similarly produces a signal at *m/z* 87. In other words, 2,3-butanedione and pentanal have the same nominal mass, i.e., they are isobaric, thus if both compounds are present together in a sample they cannot be detected individually. One approach to overcome this challenge in PTR-MS analysis is to manipulate the conditions in the drift tube (reaction chamber) to induce fragmentation in the hope that the fragment ions of the two isobaric compounds differ sufficiently enough to enable their distinct detection (*1*); a similar procedure is explicitly exploited in tandem mass spectrometry (MS/MS) by use of a dedicated collision-induced dissociation chamber to establish unique fragment ions to differentiate isobaric compounds, as has been demonstrated for APCI-MS/MS (see Chapter 7). The use of alternative reagent ions, as employed in SRI-PTR-MS (and similarly in SIFT-MS; see Chapter 1), is another method to address this conundrum. Although these approaches can be useful for samples containing only a few compounds, however, this is rarely the case for real foods. A partial solution to this difficulty is offered by PTR-TOF-MS, which first emerged as an alternative to PTR-QMS in the mid-2000s (*80*), followed by the launch of a commercial system by IONICON Analytik GmbH (Innsbruck, Austria) a few years later (*81*). The higher mass resolving power afforded by TOF-MS meant that ions could – and can – be detected with greater mass accuracy.

Returning to the previous example of the two isobaric compounds, the exact masses of protonated 2,3-butanedione and pentanal are m/z 87.044 and m/z 87.080, respectively. Accordingly, these two compounds are distinguishable by TOF-MS, provided that the mass resolving power ($m/\Delta m$) is sufficiently high; in this case, a value of approximately 2500 or greater is required to resolve the Δm of 0.036. The mass resolving power of current commercial PTR-TOF-MS instruments are in the range 1000-10000 and thus offer a suitably high degree of separation for most VOC detection scenarios, for example, $\Delta m \sim 0.015$ at a mass resolving power of 6000 (*82*). Further, the improved mass accuracy exhibited by PTR-TOF-MS aids the assignment of ion traces to compounds, since the elemental composition of a signal may be deciphered, to a degree, from the accurate m/z information (*83, 84*).

In addition to the greatly expanded analytical range made available by PTR-TOF-MS through its higher mass accuracy, the system offers an additional benefit through its rapid data acquisition of mass spectra. Whereas PTR-QMS requires tens of seconds to detect all ions within a typical mass spectral range (e.g., m/z 20-200), a mass spectrum in PTR-TOF-MS is recorded in seconds (or fractions of a second, if required). This feature is especially beneficial in nosespace analysis, which requires breath-by-breath resolution of target analytes, as well as other applications with limited sample volume and thus limited time for analysis. Thus, PTR-TOF-MS can be used to monitor the *in vivo* or *in vitro* release of tens of compounds simultaneously, compared to only a handful of compounds via PTR-QMS. Similarly, these fast acquisition rates for entire mass spectra make PTR-TOF-MS ideally suited to high-throughput analysis.

Indeed, the development of PTR-TOF-MS provided new impetus for a wave of food-flavor studies with these extended capabilities, although the benefit of high mass accuracy and the faster and more comprehensive data acquisition come with a caveat of adding complexity to data processing and interpretation (*85–87*). As mentioned above, two particularly suited applications for PTR-TOF-MS are nosespace analysis and high-throughput analysis. For the former, PTR-TOF-MS technology has been used to explore flavor release in relation to sensory perception for numerous foods, including cereal bars (*88*), coffee (*89*), wine (*76*), and chewing gum (*90*), as well as aroma persistence in the nasal and oral cavities (*37*). The second application, high-throughput analysis, has especially benefitted from the coupling of a PTR-TOF-MS instrument to a robotic automated sampling system (*91, 92*). PTR-TOF-MS technology represents the final frontier in comprehensive dynamic flavor analysis, yet the limits of its analytical capabilities can be further extended by its combination with peripheral devices.

Expanding the PTR-MS Repertoire

Alongside the many opportunities in food/flavor analysis offered by PTR-MS technology, the challenges presented are numerous. These include: assigning compounds to ion traces (see Chapter 1); accurate and rapid quantitation of individual compounds; or distinguishing between isobaric and/or isomeric species. Some of these challenges can be overcome, at least in part, by the use of peripheral equipment (93), such as the aforementioned autosampler for high-throughput measurements. As discussed above, the fast and comprehensive analytical capabilities offered by PTR-TOF-MS make it an ideal tool for rapid sample measurements and, correspondingly, for the analysis of many samples in quick succession. Combining the PTR-TOF-MS with an autosampler creates a high-efficiency screening tool in which samples – usually the headspace of solid or liquid samples in vials – can be analyzed in <1 min intervals, yielding a sample throughput of typically 60 samples per hour. This application has been envisioned and developed by Biasioli and co-workers (FEM, Italy), who have showcased it for several uses, including its implementation in characterizing roasted coffees of different geographic origin (94), following the progress of microbial metabolites during fermentation (78, 95) (and featured in Chapter 10), and examining the biogenic volatile emissions from plants (92).

The separation of many isobaric compounds is possible by PTR-TOF-MS, as discussed above. Isomeric compounds, on the other hand, cannot be readily distinguished through mass spectrometry alone, owing to these compounds exhibiting the same elemental composition, thus their protonated molecules have identical exact masses. A solution to this problem is offered by coupling a GC system with PTR-MS to establish compound separation through their differing retention in the GC capillary column. The merging of these two technologies was first demonstrated in 2005, at the time with the PTR-QMS instrument (96), but a more viable combination only became available with the development of PTR-TOF-MS, with its capability of rapid mass spectral acquisition, which is an important requirement in analyzing compounds that elute from a GC column in quick succession. In particular, the use of a fastGC system that caters for short GC separation runs of only several minutes on account of rapid heating of the column by electrical conduction (and, correspondingly, an equally fast cooling), combined with a bespoke inlet configuration that allows to switch between analysis on-line or via the fastGC system, represents an ideal extention to PTR-TOF-MS that has been successfully demonstrated for different applications (97, 98).

A particular benefit of using a fastGC – as opposed to conventional GC – is that intermittent analyses via this system to separate isomeric compounds in a sample do not greatly interrupt or compromise the dynamic measurements, depending on the processes being investigated. Further, in relation to the analytical challenge presented by samples with high ethanol content, as discussed earlier, the use of fastGC offers the additional benefit that ethanol elutes quickly from the column, the latter thereby effectively acting as an ethanol filter, whereby the volatile constituents eluting after the transient appearance of ethanol at the beginning of a fastGC run can be detected without interference (i.e., the primary ion depletion caused by the presence of high quantities of ethanol recovers rapidly in the subsequent absence of ethanol) (78, 93, 98). Despite these advantages, a shortcoming of the fastGC is its lower degree of separation of isomeric compounds compared to conventional GC on account of the shorter column length, rapid temperature program, and column properties. Nevertheless, the fastGC-PTR-TOF-MS system represents a novel – and still niche – tool with high potential in food/flavor research.

Finally, another notable addition to the PTR-MS arsenal is a calibration system that allows for the generation of bespoke calibration standards using pure compounds. The commercial liquid calibration unit (LCU), manufactured by IONICON Analytik GmbH (Innsbruck, Austria), is of particular value in PTR-MS analyses in food/flavor research due to the limited availability of certified gas standards of aroma compounds for calibration purposes. The LCU system utilizes liquid microliter pumps to convey aqueous aroma compound solutions via a nebulizer into a small oven, whereby the droplets become fully vaporized to achieve a constant but variable gas-phase concentration of the constituent aroma compounds for direct measurement by PTR-MS (*93*, *99*). In addition to providing a means to calibrate the PTR-MS instrument for individual compounds, the LCU is equally a useful tool to examine the degree of fragmention of target compounds in order to determine representative product ions. Futher, the system has been recently investigated for its use in a different application, namely to analyze the volatile composition of a liquid sample directly. Specifically, the system was fed with diluted whisky samples, which were vaporized in the oven via the nebulizer in the same manner as with regular calibration solutions. The vaporized whisky was then analyzed by PTR-TOF-MS, with selected aroma compounds quantified, allowing their liquid concentrations in the matrix of the whisky sample to be determined, as has been reported in a previous book in this series (*100*). The suitability of this latter approach to analyze other liquid samples, such as wine or beer, which contain higher quantities of non-volatiles or solids and thereby present a risk of clogging the microliter pumps of the LCU, amongst other challenges, is yet to be explored.

Conclusions

The use of PTR-MS in food/flavor research boasts a rich history spanning the past quarter of a century. Numerous research articles in the scientific literature demonstrate the many novel applications accessible with this innovative technology, foremost its ability to characterize dynamic processes or its use as a high-throughput screening tool. It is perhaps somewhat surprising, therefore, that PTR-MS is still considered to be a niche tool in the field of flavor analysis, despite its enormous potential. PTR-TOF-MS in particular represents a technology that has yet much to contribute to food science. Foremost, its full potential to explore sensory perception in relation to flavor release during consumption has yet to be reached. Further, its non-destructive, high-throughput capabilities make it well suited for implementation as a quality control tool, e.g., for authentication and/or fraud screening, spoilage/oxidation monitoring, or shelf-life testing. Moreover, the on-line and rapid data acquisition capacity of PTR-TOF-MS represent ideal features for process monitoring and control – not extensively covered in this chapter – such as roasting, brewing or fermentation. Certainly, the promise of future PTR-TOF-MS applications in food science is flavorsome.

Acknowledgments

I would like to express my gratitude to the many colleagues and collaborators with whom I have discussed and worked on food science related PTR-MS activities over the years, especially Pat Silcock and Franco Biasioli. I ackonweldge the open and engaging PTR-MS user community, active in food science and beyond, for the many informative and enjoyable debates on PTR-MS applications and other topics – sometimes into the early hours of the morning – over the past two decades; such open discussions have been the catalyst for many insightful and innovative studies involving PTR-MS technology and have pushed the boundaries of its use in food/flavor science.

References

1. Hansel, A.; Jordan, A.; Holzinger, R.; Prazeller, P.; Vogel, W.; Lindinger, W. Proton-Transfer Reaction Mass-Spectrometry - Online Trace Gas-Analysis at the ppb Level. *Int. J. Mass Spectrom. Ion Proc.* **1995**, *149/150*, 609–619.
2. Ferguson, E. E.; Fehsenfeld, F. C.; Dunkin, D. B.; Schmeltekopf, A. L.; Schjff, H. I. Laboratory studies of helium ion loss processes of interest in the ionosphere. *Planet. Space Sci.* **1964**, *12*, 1169–1171.
3. Munson, M. S. B.; Field, F. H. Chemical ionization mass spectrometry. I. General Introduction. *J. Am. Chem. Soc.* **1966**, *88*, 2621–2630.
4. McFarland, M.; Albritton, D. L.; Fehsenfeld, F. C.; Ferguson, E. E.; Schmeltekopf, A. L. Flow-drift technique for ion mobility and ion-molecule reaction rate constant measurements. I. Apparatus and mobility measurements. *J. Chem. Phys.* **1973**, *59*, 6610–6619.
5. Adams, N. G.; Smith, D. The selected ion flow tube (SIFT); A technique for studying ion-neutral reactions. *Int. J. Mass Spectrom. Ion Phys.* **1976**, *21*, 349–359.
6. Lindinger, W.; Albritton, D. L.; Fehsenfeld, F. C.; Schmeltekopf, A. L.; Ferguson, E. E. Flow–drift tube measurements of kinetic energy dependences of some exothermic proton transfer rate constants. *J. Chem. Phys.* **1975**, *62*, 3549–3553.
7. Smith, D.; Španěl, P. Ions in the terrestrial atmosphere and in interstellar clouds. *Mass Spectrom. Rev.* **1995**, *14*, 255–278.
8. Lagg, A.; Taucher, J.; Hansel, A.; Lindinger, W. Applications of proton transfer reactions to gas analysis. *Int. J. Mass Spectrom. Ion Proc.* **1994**, *134*, 55–66.
9. Španěl, P.; Pavlik, M.; Smith, D. Reactions of H3O+ and OH- ions with some organic molecules; applications to trace gas analysis in air. *Int. J. Mass Spectrom. Ion Proc.* **1995**, *145*, 177–186.
10. Sulzer, P.; Hartungen, E.; Hanel, G.; Feil, S.; Winkler, K.; Mutschlechner, P.; Haidacher, S.; Schottkowsky, R.; Gunsch, D.; Seehauser, H.; Striednig, M.; Jürschik, S.; Breiev, K.; Lanza, M.; Herbig, J.; Märk, L.; Märk, T. D.; Jordan, A. A proton transfer reaction-quadrupole interface time-of-flight mass spectrometer (PTR-QiTOF): high speed due to extreme sensitivity. *Int. J. Mass Spectrom.* **2014**, *368*, 1–5.
11. Lindinger, W. Reaction-Rate Constants in Steady-State Hollow-Cathode Discharges: Ar + H_2O Reactions. *Phys. Rev. A* **1973**, *7*, 328–333.
12. Jordan, A.; Haidacher, S.; Hanel, G.; Hartungen, E.; Herbig, J.; Märk, L.; Schottkowsky, R.; Seehauser, H.; Sulzer, P.; Märk, T. D. An online ultra-high sensitivity proton-transfer-reaction mass-spectrometer combined with switchable reagent ion capability (PTR+SRI-MS). *Int. J. Mass Spectrom.* **2009**, *286*, 32–38.
13. Lindinger, W.; Hansel, A.; Jordan, A. Proton-transfer-reaction mass spectrometry (PTR–MS): on-line monitoring of volatile organic compounds at pptv levels. *Chem. Soc. Rev.* **1998**, *27*, 347–354.
14. Lindinger, W.; Hansel, A.; Jordan, A. On-line monitoring of volatile organic compounds at pptv levels by means of proton-transfer-reaction mass spectrometry (PTR-MS) medical applications, food control and environmental research. *Int. J. Mass Spectrom. Ion Proc.* **1998**, *173*, 191–241.

15. Boschetti, A.; Biasioli, F.; van Opbergen, M.; Warneke, C.; Jordan, A.; Holzinger, R.; Prazeller, P.; Karl, T.; Hansel, A.; Lindinger, W.; Iannotta, S. PTR-MS real time monitoring of the emission of volatile organic compounds during postharvest aging of berryfruit. *Postharvest Biol. Tec.* **1999**, *17*, 143–151.

16. Aprea, E.; Biasioli, F.; Gasperi, F.; Märk, T. D.; Ruth, S. v. In vivo monitoring of strawberry flavour release from model custards: effect of texture and oral processing. *Flavour Frag. J.* **2006**, *21*, 53–58.

17. Boscaini, E.; van Ruth, S.; Biasioli, F.; Gasperi, F.; Mark, T. D. Gas Chromatography−Olfactometry (GC-O) and Proton Transfer Reaction−Mass Spectrometry (PTR-MS) Analysis of the Flavor Profile of Grana Padano, Parmigiano Reggiano, and Grana Trentino Cheeses. *J. Agric. Food Chem.* **2003**, *51*, 1782–1790.

18. Biasioli, F.; Gasperi, F.; Aprea, E.; Colato, L.; Boscaini, E.; Märk, T. D. Fingerprinting mass spectrometry by PTR-MS: heat treatment vs. pressure treatment of red orange juice--a case study. *Int. J. Mass Spectrom.* **2003**, *223–224*, 343–353.

19. Biasioli, F.; Gasperi, F.; Aprea, E.; Mott, D.; Boscaini, E.; Mayr, D.; Mark, T. D. Coupling Proton Transfer Reaction−Mass Spectrometry with Linear Discriminant Analysis: a Case Study. *J. Agric. Food Chem.* **2003**, *51*, 7227–7233.

20. Granitto, P. M.; Biasioli, F.; Aprea, E.; Mott, D.; Furlanello, C.; Märk, T. D.; Gasperi, F. Rapid and non-destructive identification of strawberry cultivars by direct PTR-MS headspace analysis and data mining techniques. *Sensor Actuat. B-Chemical* **2007**, *121*, 379–385.

21. Zini, E.; Biasioli, F.; Gasperi, F.; Mott, D.; Aprea, E.; Märk, T. D.; Patocchi, A.; Gessler, C.; Komjanc, M. QTL mapping of volatile compounds in ripe apples detected by proton transfer reaction-mass spectrometry. *Euphytica* **2005**, *145*, 269–279.

22. Aprea, E.; Biasioli, F.; Sani, G.; Cantini, C.; Mark, T. D.; Gasperi, F. Proton Transfer Reaction−Mass Spectrometry (PTR-MS) Headspace Analysis for Rapid Detection of Oxidative Alteration of Olive Oil. *J. Agric. Food Chem.* **2006**, *54*, 7635–7640.

23. Biasioli, F.; Gasperi, F.; Aprea, E.; Endrizzi, I.; Framondino, V.; Marini, F.; Mott, D.; Märk, T. D. Correlation of PTR-MS spectral fingerprints with sensory characterisation of flavour and odour profile of "Trentingrana" cheese. *Food Qual. Prefer.* **2006**, *17*, 63–75.

24. Aprea, E.; Biasioli, F.; Gasperi, F.; Mott, D.; Marini, F.; Märk, T. D. Assessment of Trentingrana cheese ageing by proton transfer reaction-mass spectrometry and chemometrics. *Int. Dairy J.* **2007**, *17*, 226–234.

25. Yeretzian, C.; Jordan, A.; Badoud, R.; Lindinger, W. From the green bean to the cup of coffee: investigating coffee roasting by on-line monitoring of volatiles. *Eur. Food. Res. Technol.* **2002**, *214*, 92–104.

26. Yeretzian, C.; Jordan, A.; Lindinger, W. Analysing the headspace of coffee by proton-transfer-reaction mass-spectrometry. *Int. J. Mass Spectrom.* **2003**, *223–224*, 115–139.

27. Karl, T.; Yeretzian, C.; Jordan, A.; Lindinger, W. Dynamic measurements of partition coefficients using proton-transfer-reaction mass spectrometry (PTR-MS). *Int. J. Mass Spectrom.* **2003**, *223–224*, 383–395.

28. Pollien, P.; Jordan, A.; Lindinger, W.; Yeretzian, C. Liquid-air partitioning of volatile compounds in coffee: dynamic measurements using proton-transfer-reaction mass spectrometry. *Int. J. Mass Spectrom.* **2003**, *228*, 69–80.

29. Mayr, D.; Märk, T.; Lindinger, W.; Brevard, H.; Yeretzian, C. Breath-by-breath analysis of banana aroma by proton transfer reaction mass spectrometry. *Int. J. Mass Spectrom.* **2003**, *223–224*, 743–756.
30. Roberts, D. D.; Pollien, P.; Antille, N.; Lindinger, C.; Yeretzian, C. Comparison of Nosespace, Headspace, and Sensory Intensity Ratings for the Evaluation of Flavor Absorption by Fat. *J. Agric. Food Chem.* **2003**, *51*, 3636–3642.
31. Pollien, P.; Lindinger, C.; Yeretzian, C.; Blank, I. Proton Transfer Reaction Mass Spectrometry, a Tool for On-Line Monitoring of Acrylamide Formation in the Headspace of Maillard Reaction Systems and Processed Food. *Anal. Chem.* **2003**, *75*, 5488–5494.
32. Gloess, A. N.; Vietri, A.; Wieland, F.; Smrke, S.; Schönbächler, B.; López, J. A. S.; Petrozzi, S.; Bongers, S.; Koziorowski, T.; Yeretzian, C. Evidence of different flavour formation dynamics by roasting coffee from different origins: On-line analysis with PTR-ToF-MS. *Int. J. Mass Spectrom.* **2014**, *365–366*, 324–337.
33. Sánchez López, J. A.; Wellinger, M.; Gloess, A. N.; Zimmermann, R.; Yeretzian, C. Extraction kinetics of coffee aroma compounds using a semi-automatic machine: On-line analysis by PTR-ToF-MS. *Int. J. Mass Spectrom.* **2016**, *401*, 22–30.
34. Sánchez-López, J. A.; Zimmermann, R.; Yeretzian, C. Insight into the Time-Resolved Extraction of Aroma Compounds during Espresso Coffee Preparation: Online Monitoring by PTR-ToF-MS. *Anal. Chem.* **2014**, *86*, 11696–11704.
35. Zanin, R. C.; Smrke, S.; Yeretzian, C.; Kurozawa, L. E.; Yamashita, F. Ultrasound-Assisted Emulsification of Roasted Coffee Oil in Complex Coacervates and Real-time Coffee Aroma Release by PTR-ToF–MS. *Food Bioprocess Technol* **2021**, *14*, 1857–1871.
36. Zanin, R. C.; Viegas, M. C.; Smrke, S.; Yeretzian, C.; Kurozawa, L. E.; Yamashita, F. The role of ultrasound-assisted emulsification of roasted coffee oil on aroma profile in spray-dried microparticles and its dynamic release by PTR-ToF–MS. *Eur. Food. Res. Technol.* **2021**, *247*, 865–878.
37. Sánchez-López, J. A.; Ziere, A.; I. F. S. Martins, S.; Zimmermann, R.; Yeretzian, C. Persistence of aroma volatiles in the oral and nasal cavities: real-time monitoring of decay rate in air exhaled through the nose and mouth. *J. Breath Res.* **2016**, *10*, 036005.
38. Buhr, K.; van Ruth, S.; Delahunty, C. Analysis of volatile flavour compounds by Proton Transfer Reaction-Mass Spectrometry: fragmentation patterns and discrimination between isobaric and isomeric compounds. *Int. J. Mass Spectrom.* **2002**, *221*, 1–7.
39. van Ruth, S.; Boscaini, E.; Mayr, D.; Pugh, J.; Posthumus, M. Evaluation of three gas chromatography and two direct mass spectrometry techniques for aroma analysis of dried red bell peppers. *Int. J. Mass Spectrom.* **2003**, *223–224*, 55–65.
40. Hansson, A.; Giannouli, P.; van Ruth, S. The influence of gel strength on aroma release from pectin gels in a model mouth and in vivo, monitored with proton-transfer-reaction mass spectrometry. *J. Agric. Food Chem.* **2003**, *51*, 4732–4740.
41. Frasnelli, J.; Ruth, S. v.; Kriukova, I.; Hummel, T. Intranasal Concentrations of Orally Administered Flavors. *Chem. Senses* **2005**, *30*, 575–582.
42. Boland, A. B.; Delahunty, C. M.; van Ruth, S. M. Influence of the texture of gelatin gels and pectin gels on strawberry flavour release and perception. *Food Chem.* **2006**, *96*, 452–460.
43. van Ruth, S. M.; Frasnelli, J.; Carbonell, L. Volatile flavour retention in food technology and during consumption: Juice and custard examples. *Food Chem.* **2008**, *106*, 1385–1392.

44. van Ruth, S. M.; Buhr, K. Influence of mastication rate on dynamic flavour release analysed by combined model mouth/proton transfer reaction-mass spectrometry. *Int. J. Mass Spectrom.* **2004**, *239*, 187–192.

45. Maçatelli, M.; Akkermans, W.; Koot, A.; Buchgraber, M.; Paterson, A.; van Ruth, S. Verification of the geographical origin of European butters using PTR-MS. *J. Food Compos. Anal.* **2009**, *22*, 169–175.

46. van Ruth, S. M.; Koot, A.; Akkermans, W.; Araghipour, N.; Rozijn, M.; Baltussen, M.; Wisthaler, A.; Märk, T. D.; Frankhuizen, R. Butter and butter oil classification by PTR-MS. *Eur. Food. Res. Technol.* **2008**, *227*, 307–317.

47. Galle, S. A.; Koot, A.; Soukoulis, C.; Cappellin, L.; Biasioli, F.; Alewijn, M.; van Ruth, S. M. Typicality and Geographical Origin Markers of Protected Origin Cheese from The Netherlands Revealed by PTR-MS. *J. Agric. Food Chem.* **2011**, *59*, 2554–2563.

48. Ruiz-Samblás, C.; Tres, A.; Koot, A.; van Ruth, S. M.; González-Casado, A.; Cuadros-Rodríguez, L. Proton transfer reaction-mass spectrometry volatile organic compound fingerprinting for monovarietal extra virgin olive oil identification. *Food Chem.* **2012**, *134*, 589–596.

49. Araghipour, N.; Colineau, J.; Koot, A.; Akkermans, W.; Rojas, J. M. M.; Beauchamp, J.; Wisthaler, A.; Märk, T. D.; Downey, G.; Guillou, C.; Mannina, L.; van Ruth, S. Geographical origin classification of olive oils by PTR-MS. *Food Chem.* **2008**, *108*, 374–383.

50. Déléris, I.; Saint-Eve, A.; Dakowski, F.; Sémon, E.; Le Quéré, J.-L.; Guillemin, H.; Souchon, I. The dynamics of aroma release during consumption of candies of different structures, and relationship with temporal perception. *Food Chem.* **2011**, *127*, 1615–1624.

51. Deuscher, Z.; Andriot, I.; Sémon, E.; Repoux, M.; Preys, S.; Roger, J.-M.; Boulanger, R.; Labouré, H.; Le Quéré, J.-L. Volatile compounds profiling by using proton transfer reaction-time of flight-mass spectrometry (PTR-ToF-MS). The case study of dark chocolates organoleptic differences. *J. Mass Spectrom.* **2019**, *54*, 92–119.

52. Le Quéré, J.-L.; Gierczynski, I.; Sémon, E. An atmospheric pressure chemical ionization–ion-trap mass spectrometer for the on-line analysis of volatile compounds in foods: a tool for linking aroma release to aroma perception. *J. Mass Spectrom.* **2014**, *49*, 918–928.

53. Pionnier, E.; Chabanet, C.; Mioche, L.; Le Quéré, J.-L.; Salles, C. 1. In Vivo Aroma Release during Eating of a Model Cheese: Relationships with Oral Parameters. *J. Agric. Food Chem.* **2004**, *52*, 557–564.

54. Mestres, M.; Moran, N.; Jordan, A.; Buettner, A. Aroma release and retronasal perception during and after consumption of flavored whey protein gels with different textures. 1. in vivo release analysis. *J. Agric. Food Chem.* **2005**, *53*, 403–409.

55. Mestres, M.; Kieffer, R.; Buettner, A. Release and perception of ethyl butanoate during and after consumption of whey protein gels: Relation between textural and physiological parameters. *J. Agric. Food Chem.* **2006**, *54*, 1814–1821.

56. Buettner, A. *Aroma release and perception during consumption of food taking into account physiological aspects*; Verlag Dr. Hut: Munich, 2007.

57. Buettner, A.; Otto, S.; Beer, A.; Mestres, M.; Schieberle, P.; Hummel, T. Dynamics of retronasal aroma perception during consumption: Cross-linking on-line breath analysis with medico-analytical tools to elucidate a complex process. *Food Chem.* **2008**, *108*, 1234–1246.

58. Beauchamp, J.; Scheibe, M.; Hummel, T.; Buettner, A. Intranasal Odorant Concentrations in Relation to Sniff Behavior. *Chem. Biodivers.* **2014**, *11*, 619–638.
59. Beauchamp, J.; Frasnelli, J.; Buettner, A.; Scheibe, M.; Hansel, A.; Hummel, T. Characterization of an olfactometer by proton-transfer-reaction mass spectrometry. *Meas. Sci. Technol.* **2010**, *21*, 025801.
60. Denzer, M.; Gailer, S.; Kern, D.; Schumm, L. P.; Thuerauf, N.; Kornhuber, J.; Buettner, A.; Beauchamp, J. Quantitative Validation of the n-Butanol Sniffin' Sticks Threshold Pens. *Chemosens. Percept.* **2014**, *7*, 91–101.
61. Siefarth, C.; Tyapkova, O.; Beauchamp, J.; Schweiggert, U.; Buettner, A.; Bader, S. Influence of polyols and bulking agents on flavor release. In *Proceedings of the 5th International Conference on PTR-MS and its Applications*, Obergurgl, Austria; Hansel, A., Dunkl, J., Eds.; Innsbruck University Press: Obergurgl, Austria, 2011; pp 116–119.
62. Tyapkova, O.; Bader-Mittermaier, S.; Schweiggert-Weisz, U.; Wurzinger, S.; Beauchamp, J.; Buettner, A. Characterisation of flavour–texture interactions in sugar-free and sugar-containing pectin gels. *Food Res. Int.* **2014**, *55*, 336–346.
63. Tyapkova, O.; Siefarth, C.; Schweiggert-Weisz, U.; Beauchamp, J.; Buettner, A.; Bader-Mittermaier, S. Flavor release from sugar-containing and sugar-free confectionary egg albumen foams. *LWT-Food Sci. Technol.* **2016**, *69*, 538–545.
64. Beauchamp, J.; Zardin, E.; Silcock, P.; Bremer, P. J. Monitoring photooxidation-induced dynamic changes in the volatile composition of extended shelf life bovine milk by PTR-MS. *J. Mass Spectrom.* **2014**, *49*, 952–958.
65. Zardin, E.; Silcock, P.; Siefarth, C.; Bremer, P. J.; Beauchamp, J. Dynamic changes in the volatiles and sensory properties of chilled milk during exposure to light. *Int. Dairy J.* **2016**, *62*, 35–38.
66. Silcock, P.; Alothman, M.; Zardin, E.; Heenan, S.; Siefarth, C.; Bremer, P. J.; Beauchamp, J. Microbially induced changes in the volatile constituents of fresh chilled pasteurised milk during storage. *Food Packaging and Shelf Life* **2014**, *2*, 81–90.
67. Zareian, M.; Böhner, N.; Loos, H. M.; Silcock, P.; Bremer, P.; Beauchamp, J. Evaluation of volatile organic compound release in modified atmosphere-packaged minced raw pork in relation to shelf-life. *Food Packaging and Shelf Life* **2018**, *18*, 51–61.
68. Zareian, M.; Tybussek, T.; Silcock, P.; Bremer, P.; Beauchamp, J.; Böhner, N. Interrelationship among myoglobin forms, lipid oxidation and protein carbonyls in minced pork packaged under modified atmosphere. *Food Packaging and Shelf Life* **2019**, *20*, 100311.
69. Schuster, L.; Franke, C.; Silcock, P.; Beauchamp, J.; Bremer, P. J. Development of a novel sample reuse approach to measure the impact of lean meat, bone and adipose tissue on the development of volatiles in vacuum-packed chilled lamb stored at 2 °C for 15 days. *Meat Science* **2018**, *145*, 31–39.
70. Franke, C.; Beauchamp, J. Real-Time Detection of Volatiles Released During Meat Spoilage: a Case Study of Modified Atmosphere-Packaged Chicken Breast Fillets Inoculated with Br. thermosphacta. *Food Anal. Methods* **2016**, *10*, 310–319.
71. Benjamin, O.; Silcock, P.; Beauchamp, J.; Buettner, A.; Everett, D. W. Tongue pressure and oral conditions affect volatile release from liquid systems in a model mouth. *J. Agric. Food Chem.* **2012**, *60*, 9918–9927.

72. Benjamin, O.; Silcock, P.; Beauchamp, J.; Buettner, A.; Everett, D. W. Volatile release and structural stability of β-lactoglobulin primary and multilayer emulsions under simulated oral conditions. *Food Chem.* **2013**, *140*, 124–134.
73. Benjamin, O.; Silcock, P.; Beauchamp, J.; Buettner, A.; Everett, D. W. Emulsifying Properties of Legume Proteins Compared to β-Lactoglobulin and Tween 20 and the Volatile Release from Oil-in-Water Emulsions. *J. Food Sci.* **2014**, *79*, E2014–E2022.
74. Lindinger, C.; Labbe, D.; Pollien, P.; Rytz, A.; Juillerat, M. A.; Yeretzian, C.; Blank, I. When Machine Tastes Coffee: Instrumental Approach To Predict the Sensory Profile of Espresso Coffee. *Anal. Chem.* **2008**, *80*, 1574–1581.
75. Boscaini, E.; Mikoviny, T.; Wisthaler, A.; Hartungen, E. v.; Märk, T. D. Characterization of wine with PTR-MS. *Int. J. Mass Spectrom.* **2004**, *239*, 215–219.
76. Sémon, E.; Arvisenet, G.; Guichard, E.; Le Quéré, J.-L. Modified proton transfer reaction mass spectrometry (PTR-MS) operating conditions for in vitro and in vivo analysis of wine aroma. *J. Mass Spectrom.* **2018**, *53*, 65–77.
77. Spitaler, R.; Araghipour, N.; Mikoviny, T.; Wisthaler, A.; Via, J. D.; Märk, T. D. PTR-MS in enology: Advances in analytics and data analysis. *Int. J. Mass Spectrom.* **2007**, *266*, 1–7.
78. Khomenko, I.; Stefanini, I.; Cappellin, L.; Cappelletti, V.; Franceschi, P.; Cavalieri, D.; Märk, T. D.; Biasioli, F. Non-invasive real time monitoring of yeast volatilome by PTR-ToF-MS. *Metabolomics* **2017**, *13*, 118.
79. Berbegal, C.; Khomenko, I.; Russo, P.; Spano, G.; Fragasso, M.; Biasioli, F.; Capozzi, V. PTR-ToF-MS for the Online Monitoring of Alcoholic Fermentation in Wine: Assessment of VOCs Variability Associated with Different Combinations of Saccharomyces/Non-Saccharomyces as a Case-Study. *Fermentation* **2020**, *6*, 55.
80. Blake, R. S.; Whyte, C.; Hughes, C. O.; Ellis, A. M.; Monks, P. S. Demonstration of proton-transfer reaction time-of-flight mass spectrometry for real-time analysis of trace volatile organic compounds. *Anal. Chem.* **2004**, *76*, 3841–3845.
81. Jordan, A.; Haidacher, S.; Hanel, G.; Hartungen, E.; Märk, L.; Seehauser, H.; Schottkowsky, R.; Sulzer, P.; Märk, T. D. A high resolution and high sensitivity proton-transfer-reaction time-of-flight mass spectrometer (PTR-TOF-MS). *Int. J. Mass Spectrom.* **2009**, *286*, 122–128.
82. Zardin, E.; Tyapkova, O.; Buettner, A.; Beauchamp, J. Performance assessment of proton-transfer-reaction time-of-flight mass spectrometry (PTR-TOF-MS) for analysis of isobaric compounds in food-flavour applications. *LWT - Food Science and Technology* **2014**, *56*, 153–160.
83. Cappellin, L.; Biasioli, F.; Fabris, A.; Schuhfried, E.; Soukoulis, C.; Märk, T. D.; Gasperi, F. Improved mass accuracy in PTR-TOF-MS: Another step towards better compound identification in PTR-MS. *Int. J. Mass Spectrom.* **2010**, *290*, 60–63.
84. Graus, M.; Müller, M.; Hansel, A. High resolution PTR-TOF: quantification and formula confirmation of VOC in real time. *J. Am. Soc. Mass Spectr.* **2010**, *21*, 1037–1044.
85. Cappellin, L.; Soukoulis, C.; Aprea, E.; Granitto, P.; Dallabetta, N.; Costa, F.; Viola, R.; Märk, T. D.; Gasperi, F.; Biasioli, F. PTR-ToF-MS and data mining methods: a new tool for fruit metabolomics. *Metabolomics* **2012**, *8*, 761–770.
86. Titzmann, T.; Graus, M.; Müller, M.; Hansel, A.; Ostermann, A. Improved peak analysis of signals based on counting systems: Illustrated for proton-transfer-reaction time-of-flight mass spectrometry. *Int. J. Mass Spectrom.* **2010**, *295*, 72–77.

87. Cappellin, L.; Biasioli, F.; Schuhfried, E.; Soukoulis, C.; Märk, T. D.; Gasperi, F. Extending the dynamic range of proton transfer reaction time-of-flight mass spectrometers by a novel dead time correction. *Rapid Commun. Mass Spectrom.* **2011**, *25*, 179–183.
88. Heenan, S.; Soukoulis, C.; Silcock, P.; Fabris, A.; Aprea, E.; Cappellin, L.; Märk, T. D.; Gasperi, F.; Biasioli, F. PTR-TOF-MS monitoring of in vitro and in vivo flavour release in cereal bars with varying sugar composition. *Food Chem.* **2012**, *131*, 477–484.
89. Romano, A.; Cappellin, L.; Ting, V.; Aprea, E.; Navarini, L.; Gasperi, F.; Biasioli, F. Nosespace analysis by PTR-ToF-MS for the characterization of food and tasters: The case study of coffee. *Int. J. Mass Spectrom.* **2014**, *365–366*, 20–27.
90. Pedrotti, M.; Spaccasassi, A.; Biasioli, F.; Fogliano, V. Ethnicity, gender and physiological parameters: Their effect on in vivo flavour release and perception during chewing gum consumption. *Food Res. Int.* **2019**, *116*, 57–70.
91. Capozzi, V.; Yener, S.; Khomenko, I.; Farneti, B.; Cappellin, L.; Gasperi, F.; Scampicchio, M.; Biasioli, F. PTR-ToF-MS Coupled with an Automated Sampling System and Tailored Data Analysis for Food Studies: Bioprocess Monitoring, Screening and Nose-space Analysis. *JoVE* **2017**, e54075.
92. Li, M.; Cappellin, L.; Xu, J.; Biasioli, F.; Varotto, C. High-throughput screening for in planta characterization of VOC biosynthetic genes by PTR-ToF-MS. *Journal of Plant Research* **2020**, *133*, 123–131.
93. Beauchamp, J.; Herbig, J. In *The Chemical Sensory Informatics of Food: Measurement, Analysis, Integration*; Guthrie, B., Beauchamp, J., Buettner, A., Lavine, B. K., Eds.; American Chemical Society: Washington, DC, 2015; Vol. 1191, pp 235–251.
94. Yener, S.; Romano, A.; Cappellin, L.; Märk, T. D.; Sánchez del Pulgar, J.; Gasperi, F.; Navarini, L.; Biasioli, F. PTR-ToF-MS characterisation of roasted coffees (C. arabica) from different geographic origins. *J. Mass Spectrom.* **2014**, *49*, 929–935.
95. Makhoul, S.; Romano, A.; Capozzi, V.; Spano, G.; Aprea, E.; Cappellin, L.; Benozzi, E.; Scampicchio, M.; Märk, T. D.; Gasperi, F.; El-Nakat, H.; Guzzo, J.; Biasioli, F. Volatile Compound Production During the Bread-Making Process: Effect of Flour, Yeast and Their Interaction. *Food Bioprocess Technol* **2015**, *8*, 1925–1937.
96. Lindinger, C.; Pollien, P.; Ali, S.; Yeretzian, C.; Blank, I.; Märk, T. Unambiguous Identification of Volatile Organic Compounds by Proton-Transfer Reaction Mass Spectrometry Coupled with GC/MS. *Anal. Chem.* **2005**, *77*, 4117–4124.
97. Materić, D.; Lanza, M.; Sulzer, P.; Herbig, J.; Bruhn, D.; Turner, C.; Mason, N.; Gauci, V. Monoterpene separation by coupling proton transfer reaction time-of-flight mass spectrometry with fastGC. *Anal. Bioanal. Chem.* **2015**, 1–7.
98. Romano, A.; Fischer, L.; Herbig, J.; Campbell-Sills, H.; Coulon, J.; Lucas, P.; Cappellin, L.; Biasioli, F. Wine analysis by FastGC proton-transfer reaction-time-of-flight-mass spectrometry. *Int. J. Mass Spectrom.* **2014**, *369*, 81–86.
99. Fischer, L.; Klinger, A.; Herbig, J.; Winkler, K.; Gutmann, R.; Hansel, A. The LCU: Versatile Trace Gas Calibration; In *Proceedings of the 6th International Conference on Proton Transfer Reaction Mass Spectrometry and its Applications*, Obergurgl, Austria; Hansel, A., Dunkl, J., Eds.; Innsbruck University Press: Obergurgl, Austria, 2013; pp 192–195.

100. Beauchamp, J.; Biberacher, S.; Gao, S. In *Sex, Smoke, and Spirits: The Role of Chemistry*; Guthrie, B., Beauchamp, J. D., Buettner, A., Toth, S., Qian, M. C., Eds.; American Chemical Society: Washington, DC, 2019; Vol. 1321, pp 117−124.

Chapter 4

SIFTing Through Flavor—Exploring Real-Time, Real-Life Processes Using SIFT-MS

Diandree Padayachee* and Vaughan S. Langford

Syft Technologies Limited, 68 Saint Asaph Street, Christchurch 8011, New Zealand
Email: diandree.padayachee@syft.com

Selected ion flow tube mass spectrometry (SIFT-MS) has made significant advances in food-flavor analysis in recent years. This review provides examples of SIFT-MS applications for food products – many of which have complex volatile profiles – and is forward-looking in terms of expanding on the future of SIFT-MS in food-flavor analysis. Highlighted here are several industry-specific applications, with a focus on quality control, real-time food processing, and packaging. Examples of quality control applications include food authenticity determination, checking consistency of food and beverage flavors, as well as detecting off-flavors and ensuring freshness has not been compromised. The importance of using automated SIFT-MS to obtain repeatable results for food testing applications is emphasized, along with the use of chemometrics, which would allow for rapid decision-making in a process environment. Real-time monitoring of processes is covered, which enables better understanding of flavor development, optimization of processes, controlling for the formation of potential carcinogens, and predicting volatile release with the assistance of kinetic modeling. Safety considerations in food processing can extend beyond just food safety – and SIFT-MS has also been successfully applied to the monitoring of compounds in the lead-up to potential combustion. Finally, it is also important that unwanted compounds do not migrate into food from packaging. With automated SIFT-MS, paper packaging can be rapidly screened for key sensory attributes, and residual monomers measured as they are emitted from plastic packaging. Furthermore, with thermal desorption-SIFT-MS, real-time emissions from polymers can be measured while simulating processing and usage conditions.

Introduction

Selected ion flow tube mass spectrometry (SIFT-MS) has come a long way since its introduction as an analytical technique in the mid-1990s (*1*). The benefits of this ultra-soft ionization technique, which have been outlined in Chapter 1, have led to SIFT-MS making significant inroads into the direct analysis market over the last decade in particular.

© 2021 American Chemical Society

The complexity of food volatile profiles has not been a limitation for SIFT-MS – supported by a steadily growing library of flavor compounds for method development and pseudo-absolute quantitation based on kinetic theory. The technique has in fact found widespread use in food analysis applications, for the analysis of both volatile organic compounds (VOCs) and inorganics, such as hydrogen sulfide and ammonia. SIFT-MS has been applied to the analysis of a diverse range of food products – from fruits, vegetables, nuts, seeds and oils, to dairy, poultry, red meat and seafood – as well as for food packaging analysis. These were comprehensively reviewed in 2019 (2) and will not be repeated here. As noted in the concluding remarks of that review, there are emerging themes for SIFT-MS that will see its wider adoption in industry. It therefore seemed apt to focus this chapter on industrial applications of SIFT-MS, and to provide examples of work done primarily in collaboration with food processing/manufacturing industry partners.

In a process environment, automation is key to ensuring product consistency. Quality control (QC) naturally implies minimizing variability, and that would also apply to any analytical tool in the quality assurance (QA)/QC process. A significant amount of work reported here was therefore carried out with automated SIFT-MS.

Automated headspace analysis was successfully achieved with the integration of commercial SIFT-MS instruments (Syft Technologies, Christchurch, New Zealand) to syringe-injection autosamplers. Briefly, the autosampler and SIFT-MS are interfaced by means of a self-sealing injection port (septumless head, GERSTEL, Mülheim an der Ruhr, Germany) that the syringe needle pushes open upon injection. This approach allows for slow, well-controlled injection of the sample headspace synchronous with the SIFT-MS analysis. The GERSTEL Multipurpose Sampler (MPS) is the most commonly used autosampler option for SIFT-MS, which also allows for intelligent scheduling of samples with the GERSTEL Maestro software. In this way, sample throughput is optimized.

To make sense of complex food flavor profiles sometimes necessitates the use of chemometrics. In an industrial setting, the statistically processed data is then used to make a decision. Examples showing the application of multivariate statistics to SIFT-MS analysis are discussed throughout this chapter – including an example of the first attempt at discrimination in food analysis using negative ions. However, before delving into the industrial applications, a basic understanding of how multivariate statistical methods have been applied to SIFT-MS data is first necessary.

Chemometrics and SIFT-MS: Turning Chemical Data into Actionable Decisions

SIFTing flavor attributes in industrial applications of SIFT-MS often focuses on matching characteristic patterns of odorants (rather than individual odorants) to, for example, confirm origin, assure quality, detect adulteration, or ensure packaging will not taint food. The most practicable means to utilize odorant patterns in decision making is to apply multivariate statistical analysis to the instrumental data. Various multivariate statistical methods have been developed for the variable-rich data obtained by chromatographic, mass spectrometric, and spectroscopic techniques (3).

SIFT-MS instruments can be operated in either untargeted (SCAN) or targeted (selected ion monitoring (SIM)) modes, with both types of data being amenable to multivariate statistical analysis. Most studies published to date have focused on sample classification and utilized soft independent modelling by class analogy (SIMCA) (4). In SIMCA, principal component analysis (PCA) is applied to the entire dataset and to each of the classes to create a model that discriminates each class from the others. Three types of output from SIMCA analysis are usually discussed in the literature (2). In brief, these are: (i) class projections, which visually represent where samples fall with respect to the three most important principal components derived from PCA on the entire dataset; (ii) interclass distances, which measure separation between classes (a value of three or greater is usually considered acceptable for class separation (5)); and, (iii) discriminating power, which identifies variables that provide the most discrimination between the classes (compounds or *m/z* for targeted and untargeted analyses, respectively).

Industrial Applications of SIFT-MS

Quality Control

Quality control of food products is a reactive but necessary process and can be carried out for various reasons in the food industry – such as confirming food authenticity, ensuring product consistency, checking that food flavor and freshness have not been compromised, or assuring that food is safe to consume.

Food authenticity issues can arise from adulteration, origins and even processing methods. This is often economically motivated (since authentic foods come with a higher price tag), but results in a loss of consumer confidence. The combination of targeted or untargeted SIFT-MS analysis with chemometrics has been very successful in food authenticity determination. One such example was being able to differentiate smoked salmon produced by traditional wood smoking, versus industrial liquid smoking – with compounds such as furfural and methyl furfural being clear indicators of the wood smoking process (6).

Plant-sourced oils with superior organoleptic, health, and/or cosmetic benefits – such as Argan oil and olive oil – are increasingly susceptible to adulteration and fraudulent origin claims due to their higher prices. Although early work on olive oil from SIFT-MS pioneer Murray McEwan's laboratory focused on aroma profiling (7) and oxidative status (8, 9), recent research from several groups in Europe has focused on evaluating SIFT-MS for rapid determination of oil origin and purity (i.e., whether it has been adulterated).

The adulteration of extra-virgin olive oils (EVOO) with seed oils was tested using targeted SIFT-MS analysis combined with multivariate statistical analysis (10). The SIMCA classification approach demonstrated that the aroma profile of EVOO and various seed oil adulterants are distinct. Adulterant concentrations were modeled using partial least squares regression (PLS-R), demonstrating an ability to detect adulteration down to 5% volume:volume.

Untargeted SIFT-MS analysis combined with multivariate statistical analysis has been demonstrated to be effective in authenticating the geographic origin of Moroccan Argan oils and virgin olive oils from various regions around the Mediterranean. This approach proved extremely effective at rapidly classifying 130 virgin olive oils from six Mediterranean growing regions compared to various conventional chemical measures (11). Similarly, SIFT-MS proved very effective and rapid for classification of 95 extra-virgin Argan oils originating from five Moroccan growing regions (12). Key advantages of using SIFT-MS in this way are (i) that no prior knowledge of volatiles is required for experimental set-up, and (ii) any variation in volatiles outside of the target compound list is also detected – a factor that can be very significant in QA/QC applications.

Whether operated in targeted or untargeted mode, SIFT-MS coupled with multivariate statistical analysis is clearly able to provide high-throughput origin and purity confirmation of high-value oils. Rapid switching between reagent ions with SIFT-MS allows for use of an alternative reagent ion – to mitigate the effect of fast-reacting reagent ions being depleted by high-concentration matrix compounds – without delays in analysis, e.g., the use of NO^+ in a high methanol environment. In instances where all reagent ions react rapidly with a compound, e.g., ethanol, sample dilution then becomes necessary. The dilution of duplicate aliquots of beer samples to an alcohol content of 0.5 % worked well for the headspace analysis of acetaldehyde and off-flavor sulfur compounds (dimethyl sulfide, dimethyl disulfide, dimethyl trisulfide, hydrogen sulfide and methional) with SIFT-MS (13). These challenging compounds for gas chromatography mass spectrometry (GC/MS) were easily and repeatably analyzed with automated SIFT-MS, despite the pre-dilution of the beer samples. Automated SIFT-MS also advantageously allows for further in-line dilution of the sample while it is being injected into the SIFT-MS instrument, by utilizing a make-up gas to ensure the standard flow into the instrument is maintained.

Another important aspect to beer quality, and currently a significant challenge in brewing science, is how well the beer ages. This is different for each beer because of the variability in raw materials and processing, and therefore requires chemometrics to handle the complex information (14). An initial investigation was carried out to determine if beer that was deliberately aged by leaving it out at room temperature in a clear glass bottle for 10 days could be distinguished from beer stored in the refrigerator (15). Six replicates of each beer type were tested in SIM mode, and the concentrations of selected beer aging compounds (14) and some additional sulfur compounds were processed with SIMCA analysis. The fresh and aged beers were clearly distinguished from each other, with acetaldehyde providing the greatest discrimination.

Quality control of beers extends beyond just measuring and controlling for the off-flavors, as consumers also expect consistency of both flavor and aroma in their choice of beverage. Flavor profiling of beer is also just as easily done with automated SIFT-MS, and this was demonstrated with a 29-compound targeted headspace analysis of two normal lagers, one non-alcoholic lager and a water blank (16). PCA was applied and showed clear separation for all samples – in particular for the non-alcoholic lager from the alcoholic beers. A loadings plot highlighted ethanol as a significant contributor to the separation between non-alcoholic and alcoholic beers, and also identified the compounds responsible for each beer's flavor profile.

For food and beverage samples with ethanol concentrations at moderate levels, analysis without any prior sample preparation is possible using the NO^+ reagent ion – which has a slightly slower rate coefficient than H_3O^+ and O_2^+ for the reaction with ethanol. In fact, the indispensability of the NO^+ reagent ion in food-flavor analysis has long been apparent to users of SIFT-MS instruments. Its unique association reaction mechanism for both ketones and esters, as well as its robustness to variations in humidity, provides both enhanced selectivity and stable quantitation in food matrices. Rapid fingerprinting and discrimination of different strawberry-flavored mixes was shown to be possible using only the NO^+ reagent ion, which was selected due to the relatively high levels of ethanol in the flavor mixes (17). Untargeted automated SIFT-MS analysis of three batches of two flavor standards (S1 and S2), as well as three unknown samples (U1, U2 and U3), were carried out. Five replicates of each sample were analyzed to allow for post-processing with multivariate statistical analysis. Analysis showed that there was significantly less inter-batch variation for S2 than S1. Furthermore, the unknown samples were correctly classified as follows (Figure 1):

- U1 was another batch of S1
- U2 was another batch of S2
- U3 was from a totally different flavor mix altogether

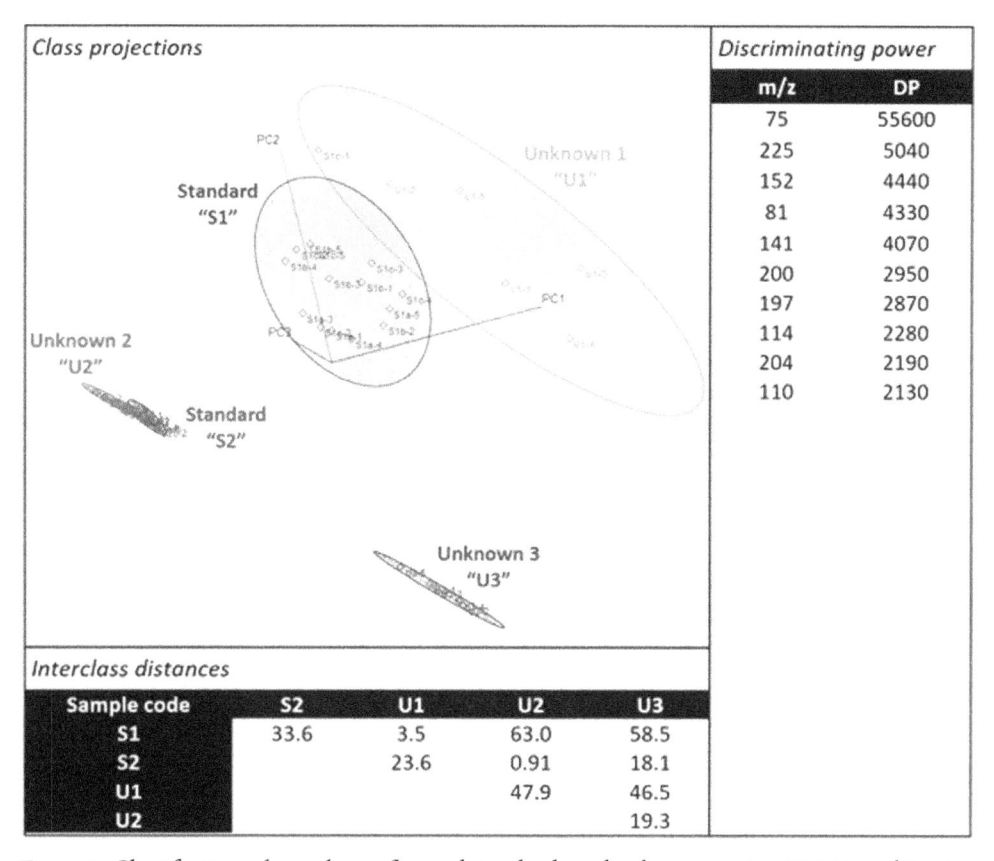

Class projections		Discriminating power	
		m/z	DP
		75	55600
		225	5040
		152	4440
		81	4330
		141	4070
		200	2950
		197	2870
		114	2280
		204	2190
		110	2130

Interclass distances				
Sample code	S2	U1	U2	U3
S1	33.6	3.5	63.0	58.5
S2		23.6	0.91	18.1
U1			47.9	46.5
U2				19.3

Figure 1. Classification of strawberry-flavored standards and unknowns using SIMCA multivariate statistical analysis coupled with SIFT-MS

Automated SIFT-MS was also successfully applied to origin testing of green coffee beans (*18*). Headspace concentrations of three batches of coffee beans of five different origins were measured. Of the 26 compounds measured, the most significant discrimination of origin was obtained from five aldehydes – benzaldehyde, hexanal, 2-methylpropanal, 2-methylbutanal and pentanal. After roasting, the aldehyde concentrations in the roasted beans were found to be significantly different to the green beans, but unlike the green beans, could not be used to discriminate between origins. This could be seen as an advantage in terms of not having to do any processing of the beans before confirming coffee origins.

Imitations of Parmigiano Reggiano cheese, or Parmesan as it is called in English-speaking parts of the world, are often rife in supermarkets outside of Europe. It was previously demonstrated that SIFT-MS can readily discriminate genuine Italian Parmesan from imitation New Zealand variants (*19*) even when the target compound list was restricted to the most significant odor-active species identified by Qian and Reineccius (*20–23*). More recently, the ability to differentiate genuine Parmesan cheese from three Italian manufacturers was tested (*24*). Ten replicate samples each of six products – of which four products were from one manufacturer (Table 1) – were analyzed in untargeted SCAN mode with automated SIFT-MS, and the signals were then reprocessed to give concentrations of the odor-active compounds.

Table 1. Manufacturer and sample information for Parmigiano Reggiano cheeses tested with SIFT-MS

Supplier and Product Name	Manufacturer Code	Sample Format	Label
Sainsbury's	08 158	Pre-grated	P1
Tesco	08 039	Wedge	P2
Aldi "Specially Selected"	08 621	Wedge	P3
Sainsbury's "Taste the Difference"	08 158	Wedge	P4
Sainsbury's	08 158	Wedge	P5
Sainsbury's "SO Organic"	08 158	Wedge	P6

Multivariate statistical analysis using SIMCA showed that all products could be distinguished from each other – i.e., there was also variability among cheeses from the same manufacturer. Evaluations of the ability to discriminate between different manufacturers showed promise, but was only moderately conclusive, due to partial overlap of some replicates of P2 and P6.

This was followed up with classification of Parmesan cheeses using untargeted analysis in both positive and negative ionization modes (25). Interestingly, this approach yielded better discrimination than when the odor-active compounds were used. Discrimination between products was successfully achieved for all three positive reagent ions (H_3O^+, NO^+ and O_2^+), but the standout discriminating ion was NO^+, for which discrimination by manufacturer was also achieved. The negative reagent ion, OH^-, also showed promise for discriminating between samples – except for not being able to completely separate P2 from P1 and P5, which in turn also affected the discrimination by manufacturer. This is, however, a promising start to further investigations of the use of negative ions in SIFT-MS food applications. Figure 2 shows the discrimination by manufacturer achieved with the NO^+ and OH^- reagent ions.

The measurement of small volatile molecules for quality control purposes is readily accomplished with SIFT-MS – of which one important application is the simultaneous analysis of ethylene and other volatile indicators of fruit ripening, for which fast analysis is a challenge for chromatography. For green kiwifruit producers, long shipping times and high shipping costs necessitates checking that their produce is at optimal ripeness before shipping. At the same time, early indication of rot or disease is also invaluable information to obtain before shipping. This was demonstrated by the easy detection of the presence of a single fruit that was over-ripe or afflicted with Botyris rot in a tray of export-quality fruit (26). Automated high-throughput analysis of fruit was demonstrated in a study comparing the measurement of small molecules from three types of apple, as well as single pear and banana varieties (27). Automation coupled with the efficient scheduling enabled by the GERSTEL Maestro software, allowed for six times higher sample throughput than GC/MS in a 24-hour period.

(a)

Class projections

(PCA plot showing P3 cluster separated along PC2; P2 cluster and P1, P4, P5, P6 cluster along PC1/PC3)

Discriminating power

m/z	DP
88	1371
59	699
69	504
43	338
45	215
74	202
133	159
44	115
116	112
89	100

Interclass distances

Manufacturer	P2	P3
P1, P4, P5, P6	4.8	15.4
P2		17.8

(b)

Class projections

Discriminating power

m/z	DP
190	1678
94	485
204	468
220	347
43	312
106	296
177	287
181	178
164	166
146	154

Interclass distances

Manufacturer	P2	P3
P1, P4, P5, P6	1.7	8.1
P2		19.3

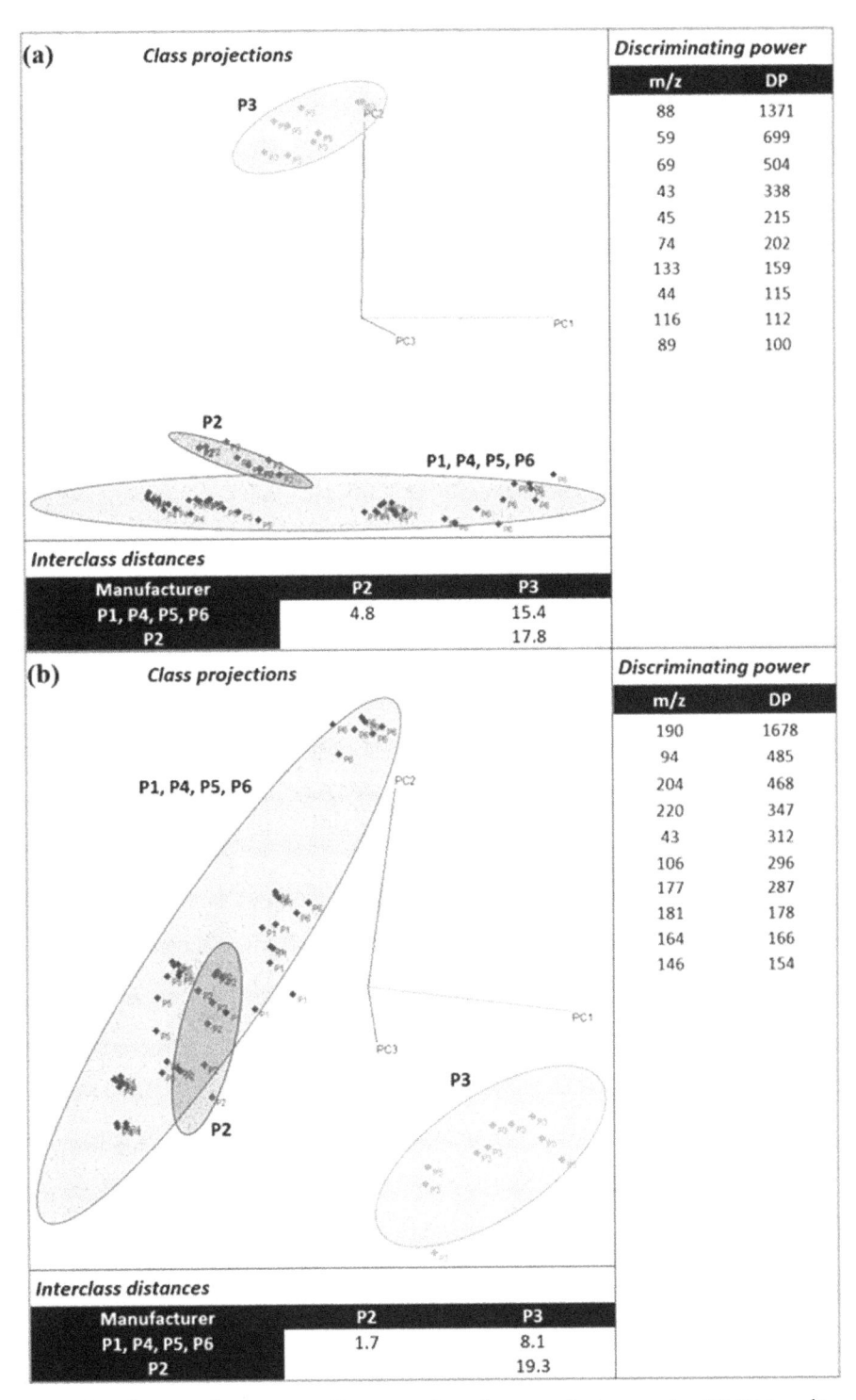

Figure 2. Classification of Parmigiano Reggiano cheese by manufacturer, using SIMCA multivariate statistical analysis coupled with SIFT-MS operating in untargeted mode. Discrimination with the (a) NO^+ reagent ion and (b) OH^- reagent ion are shown.

The ease of analysis of chromatographically challenging species – such as aldehydes, amines, reduced sulfur compounds and volatile fatty acids – combined with the significance of many of these species in both favorable and unfavorable flavors, has meant that meat flavor analysis is an ideal SIFT-MS application. The most notable industrial application of SIFT-MS to meat flavor analysis was the discrimination of samples of prime beef from beef samples with various flavor defects (*28*). SIFT-MS analysis of beef volatiles was correlated with the sensory analysis grading, using multivariate statistical analysis. There was clear discrimination between prime beef and various defective samples of beef, demonstrating the use of SIFT-MS as a fast, economical grading tool for beef – which would also allow for larger numbers of samples to be processed and increased confidence in the quality of beef products going to market.

Finished food and beverage products are sometimes stored in warehouses with no temperature regulation. Concerns from a manufacturer about the stability of their energy drinks being stored in warehouses in hot climates led to the testing of bottles from the same batch of a Thai energy drink over a 10.5-week period (*29*). Two bottles used as a control were tested, then stored in a refrigerator at 4 °C, while the remaining bottles were stored unopened at 35 °C. Freshly opened bottles of energy drinks were then analyzed in duplicate in weeks 4, 7 and 10.5, along with the control. The drinks stored at 35 °C showed significant reductions in the concentration of ethyl esters accompanied by an increase in ethanol concentrations over time, compared to the refrigerator controls. This indicated a breakdown of several flavor components of drinks stored in warm environments, resulting in reduced intensity of flavor over time. These results then enabled the manufacturer to make informed decisions on the long-term storage of their product.

Real-Time Food Processing

Obtaining greater understanding of flavor formation and development during food processing is readily done with the application of real-time SIFT-MS monitoring. Processes that can be tracked include enzymatic reactions, roasting, mixing and grinding, and this enables better product development and process optimization. However, the application of SIFT-MS in processing is not just limited to analysis of aroma compounds. In some instances, the formation of potentially harmful compounds during food processing needs to be controlled, while at other times there are occupational health and safety aspects to processing that can be better controlled for with SIFT-MS. Examples of each of these scenarios are given below.

Coffee aroma comprises a complex mixture of over 900 VOCs – of which about 5 % have sensory relevance (*30*). While commercial coffee roasting is still treated as more of an art than a science, often relying on sensory and physical measurements, there are limitations to this approach. These include the subjectivity of sensory perception, inconsistency of raw materials and inability to track flavor profiles during the roasting process. For many consumer products, consistency plays a large role in customer satisfaction, and an online objective sensory measurement would go a long way in providing that. In a proof-of-concept study (*31*), SIFT-MS was used to track the volatile emission profiles of 35 coffee aroma compounds (including pyrazines, furan derivatives, alcohols, aldehydes, ketones and organic acids) during roasting of a single coffee bean. Not only were the individual cracks of the coffee bean picked up as "peaks" in the concentration response to volatile release, but changes in volatile emissions when the roasting conditions were varied were also easily followed.

Online SIFT-MS measurements have been used to assist the predicted mechanism of volatile generation during different coffee roasting conditions (*32*). The release patterns of seven important coffee aroma compounds were measured during roasting and found to correspond to VOC formation and retention in the coffee extract. A good statistical fit of the experimental data and estimation of kinetic parameters was obtained. Furthermore, there was good agreement between predicted VOC concentrations and experimental values, which would allow for predicting VOC release for different roaster types and optimizing roasting conditions to deliver a more consistent product.

Formation of the potential carcinogen, acrylamide, is governed by legislation, such as Commission Regulation (EU) 2017/2158 (33), for the purpose of reducing the presence of acrylamide in food. Because acrylamide is formed in baked or fried carbohydrate-rich food, being able to track and therefore control its formation throughout its processing is invaluable. Acrylamide formation during food processing was monitored with SIFT-MS by measuring a process exhaust. Other volatile compounds were measured at the same time, including Maillard reaction products, which is useful for both product development and quality control purposes. In this instance, having the ability to monitor the formation of acrylamide and Maillard reaction products reduces the need for post-production testing and catches issues as they arise.

Also considered to be possible carcinogens are a number of volatile *N*-nitrosamines, some of which can be formed during heating of processed meat products (34). In 2013, the SIFT-MS library was therefore expanded to include several volatile nitrosamines. The emissions of *N*-nitrosodimethylamine (NDMA) and *N*-nitrosodiethylamine (NDEA) from salami that had been pre-cooked either over a flame grill or in a microwave oven were then measured (35). The flame-grilled meat emissions were 74 ppbv NDMA and 85 ppbv NDEA. On the other hand, the salami that had been microwaved at 1000 W for 10 s had total headspace concentrations of 120 ppbv NDMA and 40 ppbv NDEA, indicating that changes in volatile nitrosamine emissions resulting from different heat treatment conditions could be readily analyzed with SIFT-MS. This would allow food manufacturers to derive conditions under which formation of nitrosamine compounds are minimized during processing, and provide cooking guidelines for consumers.

The spray-drying of milk powder is a non-trivial affair and involves more than just quality control. There is a significant safety hazard which also has to be controlled for, in that accumulation of milk powder around spray jets in the drier can result in a series of exothermic reactions, rapidly increased temperatures and a potentially explosive combustion – which can cripple an entire facility (36). The monitoring of carbon monoxide (CO) for combustion detection has shortcomings, such as possible interference from other combustion sources and the short time interval between powder combustion detection and imminent explosion. The application of online SIFT-MS as a more reliable combustion detection device was therefore investigated. In the initial study, volatile products unique to Maillard and lipid oxidation reactions occurring in the dryer were monitored, and results showed that SIFT-MS could detect and quantify compounds indicative of the early stages of smoldering in both skim and whole milk powder (37). The NO$^+$ reagent ion proved very effective for this application, providing high sensitivity and selectivity, while easily accommodating the high-humidity of the sample stream. A follow-up study involving a direct side-by-side comparison of VOC and CO profiles during heating of milk powders, showed that multiple VOCs correlated well with CO and hence could be used as alternative indicators of the smoldering process (Figure 3) (38). VOC detection is far less susceptible to interferents in the spray drier's air intake. Furthermore, being able to track multiple independent compounds reduces the risk of false positives, making detection by this method more reliable. This is even more significant for direct-fired driers, where safety monitoring using VOCs resolves the inherent CO background issue. There is also the added advantage of using SIFT-MS for online product quality monitoring at the same time (for example, to avoid browning)

– meaning that both safety and quality aspects can be derived from the same analysis with a time resolution of a few seconds.

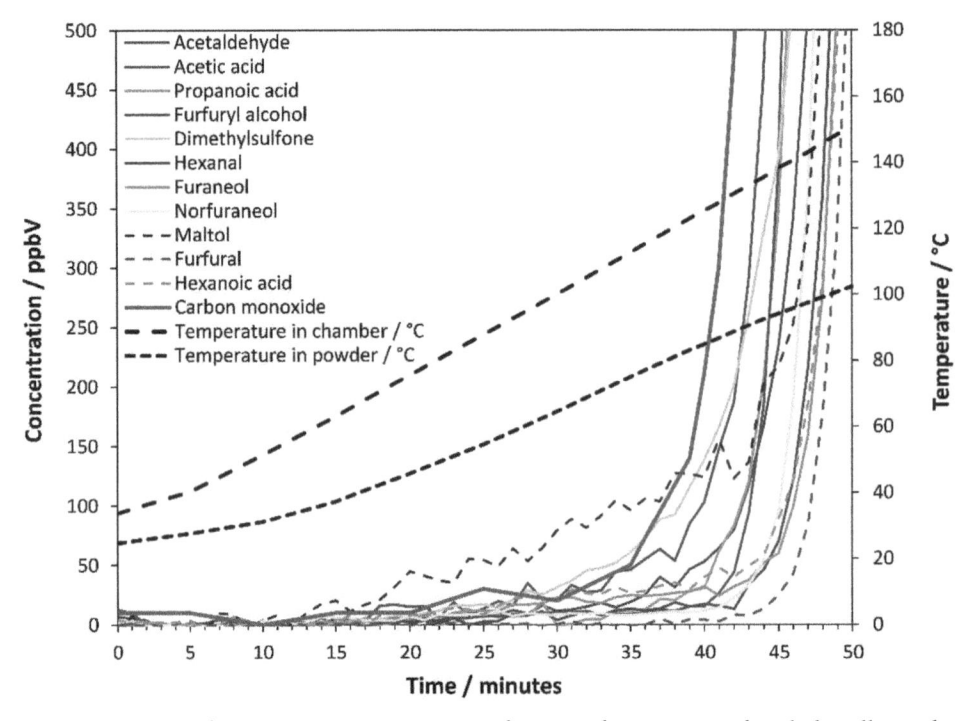

Figure 3. VOC and CO concentration traces as a function of temperature, for whole milk powder

Food Packaging

Packaging can affect both the taste and quality of food. The real-time release of VOCs from packaging raw materials and finished products is therefore of significant relevance to the food industry, particularly with the trend towards renewable packaging materials.

As a renewable packaging alternative, paper contains diverse volatile compounds arising from the source wood product, processing, as well as from the use of recycled fibers. Hence the paper packaging renaissance brings with it the need for more frequent quality control checks because there is greater variation in raw materials (especially if a proportion is recycled fiber) than for polymeric packaging materials. This means that traditional sensory methods used by industry (39–42) with high training demand, high cost per sample, and low throughput, will be unable to meet new demand. To this end, SIFT-MS was evaluated for its ability to provide instrumental prediction of odor intensity rating and odor note (43). The industrially significant conclusion of the study was that SIFT-MS has potential for sensory classification of paper samples, both for odor intensity rating and identification of sensory odor note. Odor intensity rating is the more demanding test for instrumental methods, because – within reason – multiple combinations of odorants could give an identical intensity rating on the 0-4 scale (39). Hence the successful classification of low-to-moderate and strong odor intensity ratings, but less successful in the moderately strong range (2.75 to 3.5), was a creditable and promising result. Classification based on sensory odor note was very successful, with 14 notes classified almost perfectly. The most important discrimination compounds and their relative concentrations in each paperboard sample are shown in Figure 4.

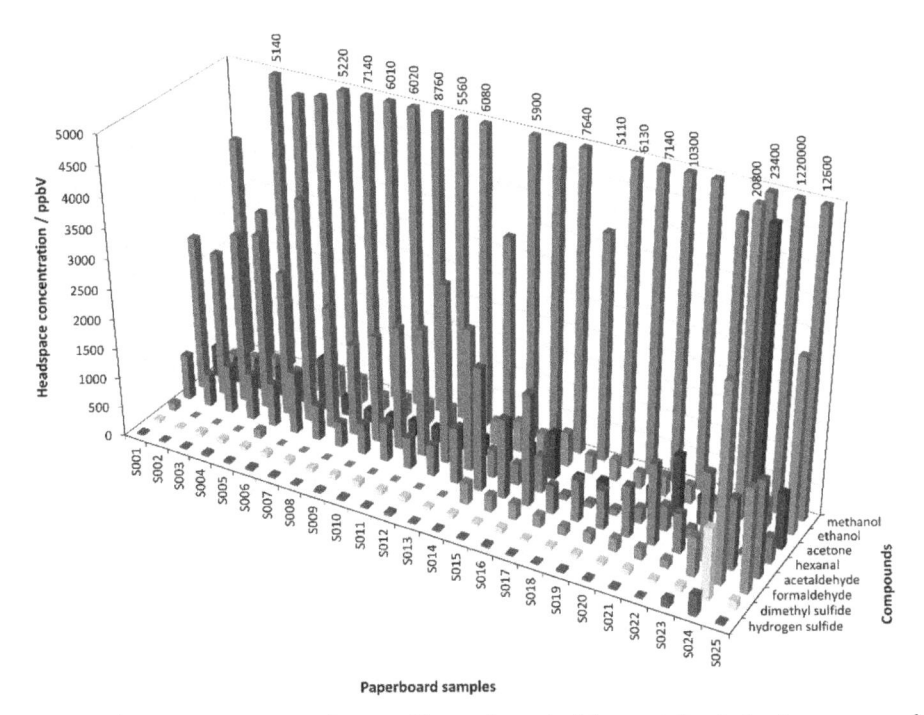

Figure 4. Headspace concentrations (mean of five replicates) of the most discriminating compounds for paper classification. Concentrations of the dominant volatile, methanol, are shown at the top of the graph for those samples where the concentration scale was exceeded

With a throughput of 12 samples per hour, SIFT-MS has potential to screen large numbers of paper samples for key sensory attributes, resolving sensory testing bottlenecks imposed by the paper packaging resurgence.

While the world transitions to renewable packaging, manufacturers of plastic packaging still have a responsibility to ensure that their products will not leach residual monomers into their contents – a most vital requirement for the food and pharmaceutical industries. Not only can automated SIFT-MS do rapid checks for residual monomer in the headspace of food packaging, but the technique has also been successfully employed to determine the total amount of residual monomer in a polymer sample using the multiple headspace extraction (MHE) technique. MHE consists of a repetition of generating headspace, measuring headspace, and flushing headspace until all the residual monomer has been removed, or sufficient data points have been gathered to enable extrapolation (typically four to six cycles). This is traditionally an expensive technique because of the slow chromatography step – a limitation resolved by using automated SIFT-MS, where a 6.5-fold increase in throughput over GC/MS was demonstrated (*44*).

Finally, it is also possible to gain insight into real-time emissions from polymers using automated thermal desorption (TD)-SIFT-MS. The earliest application of this technique was the investigation of formaldehyde emissions from two types of polyoxymethylene (POM) granules (*45*). The homo- and co-polymer forms of POM were thermally extracted at or below the recommended operating temperature or the recommended molding temperature, and the real-time formaldehyde emissions were measured with SIFT-MS. Differences in formaldehyde release behavior and concentrations for each of the two polymer forms were observed, which allowed for a rapid understanding of the stability of these polymers. This demonstrated the value of TD-SIFT-MS in testing polymer samples rapidly and easily on a small scale, while simulating processing or usage conditions.

Conclusions

Real-time SIFTing through flavor has benefits for real-life food industry. As demonstrated here, the breadth of applications being evaluated ranges from quality control to process monitoring to impurity analysis in packaging materials. This book chapter has highlighted significant steps toward process applications that are being made with the SIFT-MS technique. For ingredient, end-product, and packaging QA/QC applications in the routine testing laboratory, repeatability needs to be maximized and user input minimized in the data acquisition and reporting phases. To this end, the seamless integration of autosampler technology with SIFT-MS has been a significant step forward in recent years.

For food process applications or high-throughput screening, the traditional approach of delivering customers' target compound concentrations is inadequate. Industry needs information on which it can act immediately, and not after a scientist has processed it off-line. To this end, this chapter (and literature cited herein) has demonstrated numerous applications where SIFT-MS data have been built into multivariate statistical (i.e., chemometric) models that facilitate decision making – whether it be confirming product or ingredient origin (geographical or manufacturer), quality, freshness, or sensory acceptability. Both targeted (SIM) and untargeted (SCAN) data acquisition modes are compatible with this approach when applying classification (SIMCA) and discrimination (PLS-DA) algorithms.

Commercial SIFT-MS instrumentation now has the option of up to eight positively- and negatively-charged reagent ions. However, of those available, the three positively charged reagent ions remain the cornerstone for food applications, with NO^+ regularly performing best for both targeted applications, such as spraying drying of milk powder, and untargeted applications, such as determining the manufacturer of Parmesan or classification of strawberry flavor mixes. Three key benefits of NO^+ are (i) its relative insensitivity to variations in humidity, (ii) the multiple reaction mechanisms that enhance selectivity for carbonyl-containing compounds, in particular, and (iii) the lower sensitivity to ethanol (enabling it, for example, to be utilized in screening of flavor formulations and alcoholic beverages with only modest pre-dilution). Very limited application of the negatively charged reagent ions has currently been made to foods and flavors. Because the negatively charged reagent ions do not generally react with as many odorants as positively charged ions, it is likely that they will be utilized when positive ion options are exhausted for a given critical odorant that is detectable.

The flexibility of the SIFT-MS technique in terms of sample delivery and reporting, coupled with the stability and robustness inherent in the highly controlled chemical ionization and quadrupole-based detection, suggest that it will have a bright future in the food industry.

Acknowledgments

The authors are very grateful to Dr Mark Perkins from Anatune Limited (UK), for his contributions to several studies reported here – including analysis of samples and expertise on automated SIFT-MS.

Conflict of Interest

DP is product manager for consumer products and VSL is principal scientist at Syft Technologies Limited, a manufacturer of SIFT-MS instrumentation.

References

1. Španěl, P.; Smith, D. Selected ion flow tube: a technique for quantitative trace gas analysis of air and breath. *Med. Biol. Eng. Comput.* **1996**, *34*, 409–419; DOI: 10.1007/BF02523843.
2. Langford, V. S.; Padayachee, D.; McEwan, M. J.; Barringer, S. A. Comprehensive odorant analysis for on-line applications using selected ion flow tube mass spectrometry (SIFT-MS). *Flavour Fragr. J.* **2019**, *34* (6), 393–410; DOI: 10.1002/ffj.3516.
3. Viaene, J.; Heyden Y. V. Introduction to Herbal Fingerprinting by Chromatography. In *Chemometrics in Chromatography*; Komsta, Ł, Heyden, Y. V., Sherma, J., Eds.; CRC Press: 2018; pp 351−370.
4. Wold, S. Pattern Recognition by Means of Disjoint Principal Components Models. *Pattern Recognition* **1976**, *8* (3), 127–139; DOI: 10.1016/0031-3203(76)90014-5.
5. Kvalheim, O. M.; Karstang, T. V. SIMCA - Classification by Means of Disjoint Cross Validated Principal Components Models. In *Multivariate Pattern Recognition in Chemometrics, Illustrated by Case Studies*; Brereton, R. D., Ed.; Elsevier: 1992; pp 209–238.
6. Minaeian, J.; Ridgway, K. *The use of SIFT-MS in the assessment of good quality: smoked salmon*; Anatune Application Note AS-245; 2021.
7. Davis, B. M.; Senthilmohan, S. T.; Wilson, P. F.; McEwan, M. J. Major volatile compounds in head-space above olive oil analysed by selected ion flow tube mass spectrometry. *Rapid Commun. Mass Spectrom.* **2005**, *19*, 2272–2278; DOI: 10.1002/rcm.2056.
8. Davis, B. M.; McEwan, M. J. Determination of olive oil oxidative status by selected ion flow tube mass spectrometry. *J. Agric. Food Chem.* **2007**, *55*, 3334–3338; DOI: 10.1021/jf063610c.
9. Davis, B. M.; Senthilmohan, S. T.; McEwan, M. J. Direct determination of antioxidants in whole olive oil using the SIFT-MS-TOSC assay. *J. Am. Oil Chem. Soc.* **2011**, *88*, 785–792; DOI: 10.1007/s11746-010-1722-7.
10. Ozcan-Sinir, G. Detection of adulteration in extra virgin olive oil by selected ion flow tube mass spectrometry (SIFT-MS) and chemometrics. *Food Control* **2020**, *118*, 107433; DOI: 10.1016/j.foodcont.2020.107433.
11. Bajoub, A.; Medina-Rodriguez, S.; El-Amine, A.; Cuadros-Rodriguez, L.; Monasterio, R.; Vercammen, J.; Fernandez-Gutierrez, A.; Carrasco-Pancorbo, A. A metabolic fingerprinting approach based on selected ion flow tube mass spectrometry (SIFT-MS) and chemometrics: a reliable tool for Mediterranean origin-labelled olive oils authentication. *Food Res. Int.* **2018**, *106*, 233–242; DOI: 10.1016/j.foodres.2017.12.027.
12. Kharbach, M.; Kamal, R.; Mansouri, A. M.; Marmouzi, I.; Viaene, J.; Cherrah, Y.; Alaoui, K.; Vercammen, J.; Bouklouze, A.; Heyden, Y. V. Selected ion flow tube mass spectrometry (SIFT-MS) fingerprinting versus chemical profiling for geographic traceability of Moroccan Argan oils. *Food Chem.* **2018**, *263*, 8–17; DOI: 10.1016/j.foodchem.2018.04.059.
13. Ridgway, K. *Analysis of sulphur compounds in beer using SIFT-MS*; Anatune SnApp Note SN011; 2017.
14. Mutz, Y. S.; Rosario, D. K. A.; Conte-Junior, C. A. Insights into chemical and sensorial aspects to understand and manage beer aging using chemometrics. *Compr. Rev. Food Sci. Food Saf.* **2020**, *19* (6), 3774–3801; DOI: 10.1111/1541-4337.12642.

15. Perkins, M. J.; Padayachee, D.; Langford, V. S. *Rapid classification of beer using untargeted SIFT-MS headspace analysis*; Syft Technologies Application Note APN-065; 2021.
16. Liscio, C.; Hastie, C.; O'Connor, S. *Selected ion flow tube mass spectrometry (SIFT-MS) and chemometrics for the comparison of beer flavour profiles*; Anatune Application Note AS-206; 2018.
17. Bell, K. J. M.; Langford, V. S. Rapid determination of strawberry flavour integrity using static headspace-selected ion flow tube mass spectrometry. *Chromatogr. Today* **2019**, 48–52.
18. Khomenko, I.; Caretta, A.; Lonzarich, V.; Biasioli, F.; Navarini, L. Rapid and direct analysis of coffee beans (green and roasted) and online monitoring of coffee roasting by SIFT-MS: focus on volatile aldehydes. Poster presented at IMSC 2018, Florence, Italy.
19. Langford, V. S.; Reed, C. J.; Milligan, D. B.; McEwan, M. J.; Barringer, S. A.; Harper, W. J. Headspace analysis of Italian and New Zealand Parmesan cheeses. *J. Food Sci.* **2012**, *77* (6), C719–C726; DOI: 10.1111/j.1750-3841.2012.02730.x.
20. Qian, M.; Reineccius, G. A. Identification of aroma compounds in Parmigiano Reggiano cheese by gas chromatography/olfactometry. *J. Dairy Sci.* **2002**, *85* (6), 1362–1369; DOI: 10.3168/jds.S0022-0302(02)74202-1.
21. Qian, M.; Reineccius, G. A. Quantification of aroma compounds in Parmigiano Reggiano cheese by a dynamic headspace gas chromatography-mass spectrometry technique and calculation of odor activity values. *J. Dairy Sci.* **2003**, *86* (3), 770–776; DOI: 10.3168/jds.S0022-0302(03)73658-3.
22. Qian, M.; Reineccius, G. A. Potent aroma compounds in Parmigiano Reggiano cheese studied using a dynamic headspace (purge trap) method. *Flavour Fragr. J.* **2003**, *18* (3), 252–259; DOI: 10.1002/ffj.1194.
23. Qian, M.; Reineccius, G. A. Static headspace and aroma extract dilution analysis of Parmigiano Reggiano cheese. *J. Food Sci.* **2003**, *68* (3), 794–798; DOI: doi.org/10.1111/j.1365-2621.2003.tb08244.x.
24. Perkins, M. J.; Padayachee, D.; Langford, V. S. *Rapid Parmesan classification using automated static headspace SIFT-MS*; Syft Technologies Application Note APN-063; 2021.
25. Perkins, M. J.; Padayachee, D.; Langford, V. S. *Rapid classification of Parmesan using untargeted SIFT-MS headspace analysis*; Syft Technologies Application Note APN-064; 2021.
26. Langford, V. S.; Davis, B. M. *Rapid detection of green kiwifruit spoilage, using SIFT-MS*; Syft Technologies Application Note APN-007; 2019.
27. Perkins, M. J.; Langford, V. S. *High-throughput analysis of volatiles from fruit using SIFT-MS*; Anatune and Syft Technologies Application Note APN-030; 2018.
28. Langford, V. S.; McEwan, M. J.; Cummings, T.; Simmons, N.; Daly, C. Rapid classification of beef aroma quality using SIFT-MS. *Advances in Food and Beverage Analysis (Supplement to LCGC North America)* **2018**, 8–15.
29. Padayachee, D.; Prince, B. J. Syft Technologies Ltd. Unpublished results.
30. Toledo, P. R. A. B.; Pezza, L.; Pezza, H. R.; Toci, A. T. Relationship between the different aspects related to coffee quality and their volatile compounds. *Compr. Rev. Food Sci. Food Saf.* **2016**, *15*, 705–719; DOI: 10.1111/1541-4337.12205.
31. Gray, J.; Paterson, D.; Langford, V. S. *Process-line monitoring of coffee aroma, using SIFT-MS*; Syft Technologies Application Note APN-059; 2019.

32. Krishnaswamy, S. Kinetics of volatile generation during coffee roasting and analysis using selected ion flow tube-mass spectrometry. Dissertation. The Ohio State University, Columbus, OH, 2017.

33. European Commission Directorate-General for Health and Food Safety. European Law. *Commission Regulation (EU) 2017/2158 of 20 November 2017: Establishing Mitigation Measures and Benchmark Levels for the Reduction of the Presence of Acrylamide in Food*. https://eur-lex.europa.eu/legal-content/en/ALL/?uri=CELEX%3A32017R2158 (accessed 2021-04-14).

34. Mortensen, A.; et al. Scientific Opinion on the re-evaluation of potassium nitrite (E 249) and sodium nitrite (E 250) as food additives. *EFSA J.* **2017**, *15* (6), 4786, 1–157; DOI: 10.2903/j.efsa.2017.4786.

35. Langford, V. S.; Gray, J. D. C.; Maclagan, R. G. A. R.; Milligan, D. B.; McEwan, M. J. Real-time measurement of nitrosamines in air. *Int. J. Mass Spectrom.* **2015**, *377*, 490–495; DOI: 10.1016/j.ijms.2014.04.001.

36. Bloore, C. G.; O'Callaghan, D. Hazards in drying. In *Dairy Powders and Concentrated Products*; Tamime, A. Ed.; Wiley-Blackwell: 2009; pp 351–369.

37. Gray, J.; Davis, B.; Langford, V. Real-time monitoring of milk powder browning and oxidation products using selected ion flow tube mass spectrometry. Presented at Institute of Food Technologists 2010 Annual Meeting & Food Expo (IFT10); July 17–20, 2010; Chicago, IL.

38. Langford, V. S.; Milligan, D. B.; McEwan, M. J. Real-time monitoring of milk powder browning and oxidation. Presented at Institute of Food Technologists 2013 Annual Meeting & Food Expo (IFT13); July 13–16, 2013; Chicago, IL.

39. German Institute for Standardization. *Paper and board intended to come into contact with foodstuffs – Sensory analysis – Part 1: Odour*; DIN EN 1230-1; Berlin, 2010.

40. American Society for Testing Materials. *Standard practice for evaluating foreign odors in paper packaging*; ASTM E619-09; West Conshohocken, PA, 2009.

41. German Institute for Standardization. *Paper and board intended to come into contact with foodstuffs—sensory analysis—Part 2: Off-flavour (taint)*; DIN EN 1230-2; Berlin, 2010.

42. International Organization for Standardization. *Sensory Analysis—Methods for assessing modifications to the flavours of foodstuffs due to packaging*; ISO 13203:2003; Geneva, 2003.

43. Langford, V. S.; Du Bruyn, C.; Padayachee, D. An evaluation of selected ion flow tube mass spectrometry for rapid instrumental determination of paper type, origin and sensory attributes. *Packag. Technol. Sci.* **2021**, *34*, 245–260; DOI: 10.1002/pts.2555..

44. Perkins, M. J.; Padayachee, D.; Langford, V. S.; McEwan, M. J. High-throughput residual solvent and residual monomer analysis using selected ion flow tube mass spectrometry. *Chromatogr. Today* **2017**, 22–25.

45. Bell, K. J. M.; Ma, J.; Padayachee, D.; Langford, V. S.; Perkins, M. J. *Time-resolved thermal extraction of residual formaldehyde using thermal desorption-SIFT-MS*; Syft Technologies Application Note APN-049; 2019.

Chapter 5

Understanding the Molecular Basis of Aroma Persistence Using Real-Time Mass Spectrometry

Carolina Muñoz-González,[1,*] María Ángeles Pozo-Bayón,[1] and Francis Canon[2]

[1]Instituto de Investigación en Ciencias de la Alimentación (CIAL) CSIC-UAM, C/Nicolás Cabrera, 9, 28049 Madrid, Spain
[2]Centre des Sciences du Goût et de l'Alimentation, UMR1324 INRAe, UMR6265 CNRS Université de Bourgogne, Agrosup Dijon, F-21000 Dijon, France
*Email: c.munoz@csic.es

Aroma persistence plays a key factor driving consumers' experience during food and beverage consumption. Traditionally, aroma persistence has been studied using sensory approaches. The development of real-time mass spectrometry techniques, such as proton transfer reaction-mass spectrometry (PTR-MS) and atmospheric pressure chemical ionization-mass spectrometry (APCI-MS), has allowed us a better understanding on this complex phenomenon, highlighting the important effects of the physicochemical properties of aroma compounds on persistence. At present, the structure of aroma compounds (chemical group) has been revealed to play an important role in aroma persistence, acting in concert with human physiological factors, such as the metabolism of aroma compounds by saliva or oral cells, or the retention of aroma compounds by the salivary mucosal pellicle. Future work should be directed towards a better understanding of the effects of food matrix composition on aroma persistence, combined with the elucidation of inter-individual differences on this phenomenon. This chapter reviews the works on aroma persistence that have employed real-time mass spectrometry measurements.

Introduction

Aroma persistence – also known in the literature under other terms, such as after-smell, after-odor or after-taste – corresponds to the long-lasting perception of aromas especially after eating or drinking, but also after using other non-food products that are consumed orally, such as dental toiletries, drugs or tobacco. This phenomenon, which is a key factor in the sensory experience of consumers, is thought to be produced by a retention of aroma compounds in the consumers' breath once the product is no longer in the mouth. Thus, it is believed that aroma compounds able to remain in the mouth and throat surfaces are delivered into the air stream during breathing to reach

© 2021 American Chemical Society

the olfactory receptors over time. Hence, aroma persistence will be dependent on different factors interrelated with each other, such as the aroma compounds' structure, human physiological factors and food matrix composition. Over the last two decades, different methods have been developed to disentangle the complex phenomenon of aroma persistence through the use of real-time mass spectrometry approaches, which have allowed visualization of the dynamics of aroma compounds in the breath. This chapter provides an overview of the published empirical data on aroma persistence measured through real-time mass spectrometry techniques, as well as on the different mechanisms proposed as the origin of this phenomenon.

Exploring the Phenomenon of Aroma Persistence Using Real-Time Mass Spectrometry

The availability of real-time mass spectrometry techniques has allowed researchers to develop experiments that deliver relevant and pioneering information to understand the behavior of aroma compounds during and after the oral processing of foods and beverages (1, 2). Using such tools, it could be observed that aroma compounds released from foods or beverage products in the mouth are transported by swallowing, and by the retronasal airflow, to the olfactory epithelium located in the nose. After leaving the mouth, the first swallow (also known as 'swallow-breath') produces a high intensity of aroma compounds, but, after that, some compounds remain in the breath for a longer time. Using video-fluoroscopy and magnetic resonance imaging it was observed that, after swallowing, a viscous layer containing residues of the food product diluted in saliva was formed on the oral and oropharyngeal mucosa (3). This coating is considered one of the important factors behind aroma persistence by releasing aroma compounds over time.

Table 1 presents an overview of the scientific publications reporting on the phenomenon of aroma persistence using real-time mass spectrometry techniques. As can be seen, the first instrumental works directed to the study of aroma persistence were mostly developed around the 2000s (4–6). However, the knowledge of this complex phenomenon continues to increase and several studies have been reported in recent years (7–9). While atmospheric pressure chemical ionization-mass spectrometry (APCI-MS) has been the most employed technique, recent research has employed proton transfer reaction-mass spectrometry (PTR-MS); details of these techniques are given in Chapter 1.

To study aroma persistence, diverse parameters are extracted from the release curves. As can be seen in Table 1, however, the preferred parameter to study aroma persistence has been the ratio between the first and second (or later) breaths after swallowing. Generally, the samples analyzed have been simple aqueous solutions aromatized with single or mixtures of aroma compounds (Table 1). This has been done to minimize matrix effects. Following this premise, two papers studied aroma persistence using aroma in the vapor phase to exclude even the effect of water on the results, while focusing only on the behavior of aroma compounds in the oral and nasal cavities (4, 7). Usually, a limited number of aroma compounds (2-10) have been studied in each work, with the exception of the study by Linforth and Taylor that evaluated 41 aroma compounds (5). Matrix effects have been scarcely studied and even if some of the published studies added specific matrix components to the simple model aqueous solutions (fat, hydroxy propyl methyl cellulose), the analysis of complex and multi-composite food is rare.

Table 1. Overview of publications exploring the phenomenon of aroma persistence using real-time mass spectrometry techniques.

Reference	Technique	Parameters studied	Sample type	Aroma compounds studied	Number of panelists
Brauss et al., 1999 (10)	APCI-MS	In-nose concentration, maximum intensity (Imax)	Low and high-fat biscuits	2	3
Brauss et al., 1999 (11)	APCI-MS	time to first perception of flavor (T0), time to maximum intensity (Tmax), time for overall perception (Tend), and how long the maximum intensity lasted (Tplat).	Low and high-fat yogurt	3	1
Linforth & Taylor, 2000 (5)	APCI-MS	Ratio of breath volatile concentration for the first and second exhalations after swallowing	Aqueous solutions individually aromatized, with or without hydroxy propyl methyl cellulose (HPMC)	Up to 41 depending on the experiment	Up to 6 (3 men and 3 women, aged between 22 and 40 y/o) depending on the experiment
Wright et al., 2003 (12)	APCI-MS	Ratio between second and first breath after sample swallowing	Aqueous solutions	5	6 (3 men and 3 women, 22 and 40 y/o)
Buffo et al., 2004 (6)	APCI-MS	Ratio between the first and second breath	Aqueous solutions (aroma mixture)	5	5
Hodgson et al., 2005 (4)	APCI-MS	Decay exponent	Volatiles in gas phase or in aqueous solutions with or without HPMC; emulsions	Up to 10 depending on the experiment	Up to 5 depending on the experiment
Sanchez-López et al., 2016 (7)	PTR-MS	Decay rate	Volatiles in gas phase (aroma mixture)	8	Up to 5 depending on the experiment
Genovese et al., 2019 (13)	APCI-MS	Area under the curve (AUC), Imax	20% (v/v) olive oil-in-water (O/W) emulsion	4	2 (both men, 30 and 43 y/o)

Table 1. (Continued). Overview of publications exploring the phenomenon of aroma persistence using real-time mass spectrometry techniques.

Reference	Technique	Parameters studied	Sample type	Aroma compounds studied	Number of panelists
Muñoz-González et al., 2019 (8)	PTR-MS	AUC of the second (or later swallows) with respect to the first swallow	Wines (aroma mixture) with different tannin extracts added	5	9 (3 men and 6 women)
Muñoz-González et al., 2021 (9)	PTR-MS	AUC of the second (or later swallows) with respect to the first swallow	Aqueous solutions (aroma mixture)	5	54 (29 women and 25 men, 74 y/o)

Only four publications have evaluated aroma persistence on complex food matrices, such as biscuit, yogurts, emulsions or wine (8, 10, 11, 13). A limited number of panelists (n=1-9) participated in most of the studies, with the exception of the work from Muñoz-González and co-workers, who evaluated aroma persistence in 54 subjects (9).

Results from these studies found large differences on aroma persistence between aroma compounds, with the general observation that some compounds decayed more slowly in the panelists' breath than others. In this regard, it has been reported that the terpene alcohol (linalool) or the C_{13}-norisoprenoid (β-ionone) are present in the breath at 40% of the initial oral concentration 4 min after the oral passage of wine, while esters (isoamyl acetate, ethyl hexanoate) had almost disappeared (<10%) within this period (8). Apart from the speed of disappearance in the breath, it was also found that persistent compounds presented differences in the shape of the peak produced by the first exhalation after swallowing (5). Low-persistence compounds showed a sharp peak at the start of the exhalation followed by a small shoulder at the base of the peak, while persistent compounds had much larger shoulders relative to the maximum volatile intensity. Linforth and Taylor attributed these differences to the way in which compounds were transported from the throat, through the upper airways, and out through the nostrils (5). Moreover, some of these works (4, 6, 7) have evaluated the role of the concentration of volatiles on aroma persistence with contradictory results. While Buffo and coworkers did not observe an effect of volatile concentration on persistence (6), Sanchez and collaborators found that volatile concentration affected aroma compounds with high volatility when present at low concentration (7), and Hodgson et al. stated that under their experimental conditions, persistence was dependent on the concentration of volatile compounds in the sample (4).

Factors Underpinning Aroma Persistence

The Role of Physicochemical Properties and Structure of Aroma Compounds on Aroma Persistence

Different authors have suggested that aroma persistence is mainly driven by the physicochemical properties of aroma compounds (4–7). In 2000, Linforth and Taylor analyzed the persistence of 41 compounds in water and applied a quantitative structure property relationship approach to model aroma persistence (5). They found that hydrophobicity and volatility were the most important factors driving persistence. Other attempts at modeling aroma persistence have reinforced the idea that persistence is mostly dependent on volatility and polarity of aroma compounds (4, 7, 12), or only on volatility (6).

This hypothesis does not apply to every situation, however, and, as a result, other factors than hydrophobicity and volatility have been suggested to play a role in aroma persistence (8, 9, 14). For example, a recent study proposed that the structure of aroma compounds is an important factor to understand aroma persistence (9). Thus, the presence of specific chemical groups could drive the behavior of volatiles during consumption. A potential explanation from recent reports is that some chemical groups are susceptible to being metabolized during consumption, which would explain why some compounds, such as esters and aldehydes, have been reported in most of the works as less persistent than, for example, pyrazines or alcohols (4, 6, 8, 9) despite having similar volatility and hydrophobicity properties. Nonetheless, within the same chemical family, the physicochemical properties of aroma compounds are important factors that would explain aroma persistence.

The Role of Physiological Factors on Aroma Persistence

Although some authors have suggested that physiological factors have little influence on aroma persistence (7), this phenomenon cannot be understood without considering certain human physiological aspects, since it arises from a retention of aroma compounds in the oral and oropharyngeal cavities. In this regard, saliva has been considered a factor that can drive aroma persistence by controlling the dilution of aroma compounds in the mouth, their adsorption onto the mouth surfaces, but also the metabolism of aroma compounds in the mouth and their interaction with food matrix components (8, 9). Recently, a study pointed out two physiological mechanisms at the origin of aroma persistence (9) that are summarized in Figure 1.

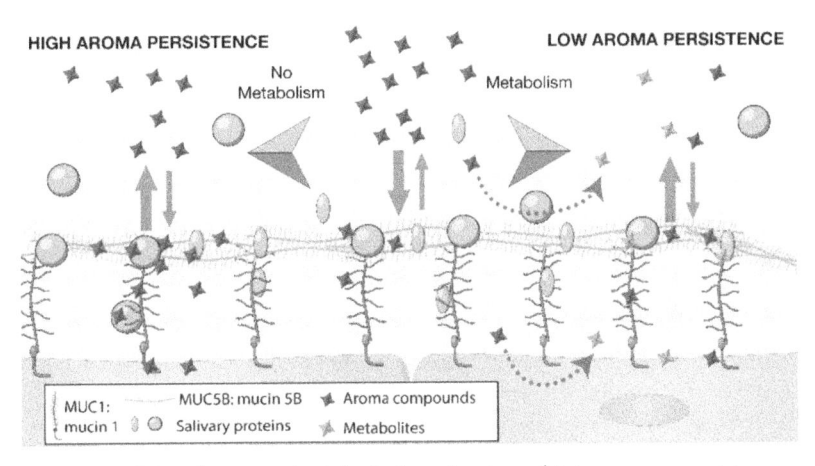

Figure 1. Hypothesis on physiological mechanisms driving aroma persistence

The first proposed mechanism is the ability of aroma compounds to interact with the mucosal pellicle. The mucosal pellicle is a hydrated layer anchored onto oral epithelial due to covalent and non-covalent bonds (15, 16). It is believed that the mucosal pellicle participate in the phenomenon of aroma persistence by retention of aroma compounds in the mouth (9). Thus, aroma compounds could interact with salivary proteins from the mucosal pellicle, such as mucins (MUC1 and MUC5B) or be dissolved in the water phase as a function of their physicochemical properties. Accordingly, aroma compounds able to interact with the mucosal pellicle would be more retained in the mouth and thus, they will show a higher aroma persistence than non-retained compounds. In this regard, it has been suggested that the retention of aroma compounds in the oral mucosa is higher than in the nasal mucosa (7). Nevertheless, it is important to bear in mind that the mucosal pellicle is constantly being replaced (17). Thus, the effect of the mucosal pellicle on aroma persistence will be dependent on how stable the mucosal pellicle is, which would be affected by the process of food intake, the presence of abrasive substances, saliva clearance or the regular desquamation of the tissue (17).

The second proposed mechanism is related to the metabolism of some aroma compounds (such as those with a carbonyl group) by saliva and oral cells (9). Accordingly, it has been shown that some aroma compounds that were subjected to metabolism in the presence of saliva and oral cells were less persistent in the breath than non-metabolizing compounds (Figure 1), as measured by both *ex vivo* and *in vivo* (PTR-MS and sensory) approaches (9). Consequently, metabolism of aroma compounds occurring *in vivo* due to the action of physiological factors (18) must be taken into account to fully understand the phenomenon of aroma persistence, as previously proposed (14, 19–21). This metabolism can be attributed to the action of salivary or cell enzymes, but also to the metabolism of aroma compounds by oral microbiota, mechanisms that will need to be explored in the future.

Finally, the effect of inter-individual differences on aroma persistence has been little explored. Although, in general, the role of inter-individual differences on aroma persistence is considered rather small (5–7), most of the reported work has been done with a limited number of individuals (Table 1). Interestingly, in a recent work using nine panelists, strong inter-individual differences on aroma persistence were found after wine consumption that were positively related to the salivary total protein content and negatively to the salivary flow of the participants for specific aroma compounds (8). This could indicate that differences in salivary parameters among individuals could be a factor affecting aroma persistence during food consumption. Nonetheless, more work in this area is necessary in order to obtain more straightforward conclusions on the effect of inter-individual differences on aroma persistence.

The Role of Food Matrix Composition on Aroma Persistence

As aroma persistence generally involves the passage of a complex food matrix through the oral cavity, it is important to consider the characteristics of the matrix to fully understand this phenomenon. On the one hand, the composition of the food matrix will govern the partitioning of aroma compounds between the product and the oral gas phase. Moreover, oral processing, such as salivation, tongue movements or chewing, will disrupt the food matrix (depending on its composition and state), affecting the release of aroma compounds to the exhaled airflow. On the other hand, the coating layer containing residues of the food product on the oral and throat surfaces and their degree of adhesion will also depend on the composition of food. In this regard, Camacho and coworkers found that the coating formed on the anterior tongue surface increased linearly with the oil content of an oil and water emulsion, with more coating resulting in higher concentration of lipophilic aroma compounds (22).

Research on the effects of matrix composition on aroma persistence are scarce. In 1999, Brauss and coworkers initiated these studies using real-time mass spectrometry techniques to determine the effect of fat content on aroma persistence using model biscuits and yogurts (Table 1) (10, 11). Overall, they found that fat content tended to increase aroma persistence. Later on, and using the hydrocolloid hydroxy propyl methyl cellulose (HPMC), several authors found a limited or no effect on aroma persistence (*4, 5*). Apart from that, Hodgson and coworkers showed that the lipid content of a beverage affected the persistence of the lipophilic compounds due to changes in the air-liquid partition coefficients of aroma compounds (*4*). To the best of our knowledge, only two works have investigated the effect of phenolic compounds on aroma persistence using real-time mass spectrometry techniques, one using emulsions (*13*) and the other one using wine (*8*) as a complex matrix. In emulsions, an effect of phenolic compounds on persistence was found to affect linalool, while in wine, the effect of phenolic compounds was only observed for the compound ethyl decanoate. In wine, additional studies carried out with alternative techniques have explored *in vivo* aroma release from the mouth of panelists. Specifically, off-line analytical approaches, such as the intra-oral solid phase microextraction (SPME) procedure (*23*) or the in-mouth headspace sorptive extraction method (*24*) have studied the effect of matrix composition (phenolic content, ethanol content) on wine aroma persistence, with interesting findings. Consequently, more studies directed towards evaluating the role of different food and beverage matrix components at different concentrations by real-time mass spectrometry approaches should be assayed in the future.

Conclusions

Aroma persistence is a complex phenomenon that has been partly disentangled in recent years thanks to the use of real-time mass spectrometry approaches, foremost APCI-MS and PTR-MS, combined with sensory experiments and the use of other biology-related techniques. These works have delivered observations that some aroma compounds are more persistent in the breath than others. The different persistence could be explained by several factors. Traditionally, it has been stated that the persistence of aroma compounds in the oral and nasal cavities is mainly dependent on the physicochemical properties of the volatile compounds; for example, compounds with high volatility and lower water solubility generally shows a low persistence in the breath. Nonetheless, new evidence suggests that the functional group of the compounds is also important to explain aroma persistence, since some chemical families, such as aldehydes and esters, can be subjected to metabolism during food and beverage consumption. Thus, compounds metabolized during consumption will be less persistent in the breath while non-metabolized compounds that are able to be retained in the oral mucosa will be more persistent. Although most of the studies have considered that inter-individual differences have a small effect on aroma persistence, they have been conducted with a limited number of panelists and without considering matrix effects that could influence the outcomes. Thus, the presence of certain matrix components, such as fat, ethanol, polyphenol content, or other factors, such as food pH or food temperature, could be important to modulate aroma persistence. To fully understand the phenomenon of aroma persistence, new multidisciplinary studies that combine real-time monitoring of aroma with sensory studies and include multi-sensory integration in the brain using a high number of panelists and different matrix types will be needed.

Acknowledgments

The authors acknowledge funding within the projects 2019T1/BIO13748 and PID2019-111734RB-100.

References

1. Lindinger, W.; Hansel, A.; Jordan, A. On-Line Monitoring of Volatile Organic Compounds at pptv Levels by Means of Proton-Transfer-Reaction Mass Spectrometry (PTR-MS) Medical

Applications, Food Control and Environmental Research. *Int. J. Mass Spectrom. Ion Process.* **1998**, *173* (3), 191–241.

2. Linforth, R.; Taylor, A. J. *Apparatus and Methods for the Analysis of Trace Constituents of Gases.* European patent EP0819937A2, 1999.
3. Buettner, A.; Otto, S.; Beer, A.; Mestres, M.; Schieberle, P.; Hummel, T. Dynamics of Retronasal Aroma Perception during Consumption: Cross-Linking on-Line Breath Analysis with Medico-Analytical Tools to Elucidate a Complex Process. *Food Chem.* **2008**, *108* (4), 1234–1246.
4. Hodgson, M.; Parker, A.; Linforth, R. S. T.; Taylor, A. J. In Vivo Studies on the Long-Term Persistence of Volatiles in the Breath. *Flavour Fragr. J.* **2004**, *19* (6), 470–475.
5. Linforth, R.; Taylor, A. J. Persistence of Volatile Compounds in the Breath after Their Consumption in Aqueous Solutions. *J. Agric. Food Chem.* **2000**, *48* (11), 5419–5423.
6. Buffo, R. A.; Rapp, J. A.; Krick, T.; Reineccius, G. A. Persistence of Aroma Compounds in Human Breath after Consuming an Aqueous Model Aroma Mixture. *Food Chem.* **2005**, *89* (1), 103–108.
7. Sánchez-López, J. A.; Ziere, A.; Martins, S. I. F. S.; Zimmermann, R.; Yeretzian, C. Persistence of Aroma Volatiles in the Oral and Nasal Cavities: Real-Time Monitoring of Decay Rate in Air Exhaled through the Nose and Mouth. *J. Breath Res.* **2016**, *10* (3), 36005.
8. Muñoz-González, C.; Canon, F.; Feron, G.; Guichard, E.; Pozo-Bayón, M. A. Assessment Wine Aroma Persistence by Using an in Vivo PTR-TOF-MS Approach and Its Relationship with Salivary Parameters. *Molecules* **2019**, *24* (7)
9. Muñoz-González, C.; Brulé, M.; Martin, C.; Feron, G.; Canon, F. Molecular Mechanisms of Aroma Persistence: From Noncovalent Interactions between Aroma Compounds and Oral Mucosa to Metabolization of Aroma Compounds by Saliva and Oral Cells. *ChemRxiv*, 2021.
10. Brauss, M. S.; Linforth, R. S. T.; Cayeux, I.; Harvey, B.; Taylor, A. J. Altering the Fat Content Affects Flavor Release in a Model Yogurt System. *J. Agric. Food Chem.* **1999**, *47* (5), 2055–2059.
11. Brauss, M. S.; Bjron, B.; Linforth, R. S. T.; Avison, S.; Taylor, A. J. Fat Content, Baking Time, Hydration and Temperature Affect Flavour Release from Biscuits in Model-mouth and Real Systems. *Flavour Fragr. J.* **1999**, *14*, 351–357.
12. Wright, K. M.; Hills, B. P.; Hollowood, T. A.; Linforth, R. S. T.; Taylor, A. J. Persistence Effects in Flavour Release from Liquids in the Mouth. *Int. J. Food Sci. Technol.* **2003**, *38* (3), 343–350.
13. Genovese, A.; Caporaso, N.; di Bari, V.; Yang, N.; Fisk, I. Effect of Olive Oil Phenolic Compounds on the Aroma Release and Persistence from O/W Emulsion Analysed in Vivo by APCI-MS. *Food Res. Int.* **2019**, *126*, 108686.
14. Buettner, A.; Mestres, M. Investigation of the Retronasal Perception of Strawberry Aroma Aftersmell Depending on Matrix Composition. *J. Agric. Food Chem.* **2005**, *53* (5), 1661–1669.
15. Bradway, S. D.; Bergey, E. J.; Scannapieco, F. A.; Ramasubbu, N.; Zawacki, S.; Levine, M. J. Formation of Salivary-Mucosal Pellicle: The Role of Transglutaminase. *Biochem. J.* **1992**, *284* (2), 557–564.

16. Gibbins, H. L.; Yakubov, G. E.; Proctor, G. B.; Wilson, S.; Carpenter, G. H. What Interactions Drive the Salivary Mucosal Pellicle Formation? *Colloids Surfaces B Biointerfaces* **2014**, *120*, 184–192.
17. Hannig, C.; Hannig, M.; Kensche, A.; Carpenter, G. The Mucosal Pellicle – An Underestimated Factor in Oral Physiology. *Arch. Oral Biol.* **2017**, *80*, 144–152.
18. Ijichi, C.; Wakabayashi, H.; Sugiyama, S.; Ihara, Y.; Nogi, Y.; Nagashima, A.; Ihara, S.; Niimura, Y.; Shimizu, Y.; Kondo, K.; Touhara, K. Metabolism of Odorant Molecules in Human Nasal/Oral Cavity Affects the Odorant Perception. *Chem. Senses* **2019**, *44* (7), 465–481.
19. Buettner, A. Influence of Human Saliva on Odorant Concentrations. 2. Aldehydes, Alcohols, 3-Alkyl-2-Methoxypyrazines, Methoxyphenols, and 3-Hydroxy-4,5-Dimethyl-2(5H)-Furanone. *J. Agric. Food Chem.* **2002**, *50* (24), 7105–7110.
20. Buettner, A. Influence of Human Salivary Enzymes on Odorant Concentration Changes Occurring in Vivo. 1. Esters and Thiols. *J. Agric. Food Chem.* **2002**, *50* (11), 3283–3289.
21. Buettner, A. Investigation of Potent Odorants and Afterodor Development in Two Chardonnay Wines Using the Buccal Odor Screening System (BOSS). *J. Agric. Food Chem.* **2004**, *52* (8), 2339–2346.
22. Camacho, S.; Van Riel, V.; De Graaf, C.; Van De Velde, F.; Stieger, M. Physical and Sensory Characterizations of Oral Coatings of Oil/Water Emulsions. *J. Agric. Food Chem.* **2014**, *62* (25), 5789–5795.
23. Muñoz-González, C.; Pérez-Jiménez, M.; Pozo-Bayón, M. Á. Oral Persistence of Esters Is Affected by Wine Matrix Composition. *Food Res. Int.* **2020**, *135*, 109286–109286.
24. Pérez-Jiménez, M.; Muñoz-González, C.; Pozo-Bayón, M. Á. Oral Release Behavior of Wine Aroma Compounds by Using In-Mouth Headspace Sorptive Extraction (HSSE) Method. *Foods* **2021**, *10* (2), 415.

Chapter 6

Using APCI-MS to Study the Dynamics of Odor Binding under Simulated Peri-Receptor Conditions

Andrew J. Taylor[*,1] and Masayuki Yabuki[2]

[1]Flavometrix Ltd., Loughborough, Leicestershire, United Kingdom
[2]Kao Corporation, 2-1-3 Bunka, Sumida-ku, Tokyo, Japan
*Email: flavometrix@btconnect.com

Odor binding protein (OBP) has been found in the olfactory systems of many insect and mammalian species. It binds a broad range of volatile organic compounds (VOCs) and one of its proposed functions in odor perception is to solubilize hydrophobic VOCs in the mucosal layer of the olfactory cleft and assist their transfer from the air phase to the olfactory receptors. The relative binding power of OBP for different volatile compounds has been measured experimentally, with values expressed as the dissociation constant, but dynamic studies are scarce. To address the lack of information, *in vitro* systems were built that could measure the release kinetics from OBP using atmospheric pressure chemical ionization-mass spectrometry (APCI-MS) as well as determine odor uptake and release by OBP under simulated *in vivo* conditions using direct mass spectrometry of gas-phase odors. The *in vitro* system was capable of introducing VOCs at a constant rate, or as pulses (to mimic the odor delivery on a breath-by-breath basis). Competition between VOCs for the binding sites on OBP was also studied experimentally.

The Role of Odor Binding Protein in Odor Perception

The overall process of olfaction occurs in several, distinct stages (*1*). In Chapter 1, the sequence of events that transfers volatile organic compounds (VOCs) from the food to the air in the mouth during oral processing was described and the value of using direct mass spectrometry (MS) to measure the dynamic nature of odor release was explained. After odor release in the mouth during eating, VOCs are transported retronasally into the nasal cavity and pass over the olfactory cleft, a small area of tissue located high in the nasal passages, where the olfactory receptors (ORs) are located. Interaction between VOCs and ORs leads to the activation of specific receptors that send signals to the brain for further processing to form the overall sense of odor perception (*2*). The area in which these interactions occur is often called the peri-receptor region (*3*). During the breathing

© 2021 American Chemical Society

cycle in humans, there is a tidal flow of air from the nose to the lungs (inspiration), followed by a reverse flow of air from the lungs that picks up air from the mouth and exits through the nose (expiration). The process of inspiration and expiration means that the ORs are alternately exposed to orthonasal and retronasal flows, respectively. During eating or drinking, this results in a pulsed delivery of odorants to the olfactory receptors.

For receptors to be activated, VOCs need to transfer from the air phase through a mucosal layer to the receptors themselves. It was originally assumed that this involved solubilization of VOCs in the mucus and subsequent uptake by the receptors. Since many odorous VOCs are hydrophobic, and sparingly soluble in aqueous media, this suggests that water-soluble odors would be more rapidly perceived than hydrophobic odors. This is not the case, and experimental data demonstrated that intranasal administration of two very water-soluble odors (propionic acid and butanol), and a very insoluble, hydrophobic odor (limonene), were perceived within similar periods (4). In that experiment, odor stimuli were delivered from an olfactometer to a human sensory panel (n=42-93) and the time between the odor being introduced into the nose and the recognition of the odor was noted. Table 1 shows the response times for the three compounds from two separate sensory sessions (4), as well as the calculated aqueous solubility and hydrophobicity values of the compounds. There was some variation in response times between sensory sessions but, overall, the range of values for all three compounds (709 to 817 ms) indicated that the hydrophilic and hydrophobic compounds were all perceived in a similar time frame despite their very different solubility and hydrophobicity values. Limonene is very difficult to solubilize in water, so the fact that it is recognized equally as quickly as the highly water-soluble compounds suggests that an alternative mechanism than dissolution in the mucus layer must exist to transfer odors from the air to the ORs. This supports the hypothesis that odor binding proteins (OBPs) are involved in the transfer of hydrophobic VOCs, a process that must involve rapid binding and release (2, 5).

Table 1. Solubility and hydrophobicity values and response times for the recognition of odors after intranasal administration (4)

Compound	Water solubility[a] (mg/L)	Hydrophobicity[a] LogP	Mean response time (ms ± SE)
Propionic acid	1.76×10^5	0.33	709 ± 15.6
1-Butanol	1.25×10^5	0.84	817 ± 14.9
Limonene	4.58×10^1	4.57	801 ± 16.9

[a] Calculated using US EPA Estimation Programs Interface Suite™ for Windows, v 4.11.

Isothermal titration studies have measured the binding of a range of VOCs to OBP, which indicates the relative affinities of the VOCs for OBP (6–8) but gives no information on the speed of binding or release. Preliminary experimental and hypothetical comparisons to evaluate the role of OBP in the olfaction process have also been published (9, 10), but ultimately, the goal of the present study was to measure the binding and release of odorants from OBP in real time and under conditions that mimicked the situation *in vivo*. It was obvious that, due to the speed of analysis and the sensitivity needed, direct injection mass spectrometry (DIMS) was the ideal technique to measure binding and release of volatile ligands by monitoring the concentration in the gas phase. The following sections describe the construction and testing of two *in vitro* systems using APCI-MS to monitor changes in VOC concentrations. The basic concept was to form a thin aqueous film of OBP on a support (to mimic the mucus layer of the olfactory epithelium), then assess the release and binding behavior of VOCs from this film by monitoring their concentration in the gas phase. By introducing VOCs in different ways, it should be possible to determine binding ratios (and compare these values with published data in order to validate the system), to mimic breathing using sequences of pulses, or to study competition between VOCs for sites on the OBP molecule by alternating the pulses between two VOCs of interest.

Developing *in vitro* Models to Explore Odorant Interactions with OBP

Off-Rate Measurements

In an attempt to measure the "off-rate", i.e., the release of odors already bound to OBP, a cellulose support was used to hold an OBP solution (with a bound VOC) in a narrow tube. Clean air was passed over the support, with the outflow connected to APCI-MS to measure the rate of odor release (9). Rat OBP-3 (11) was used in these proof-of-principle studies because the human OBP available at the time showed some instability. A solution of OBP, that had been infused with n-butyl acetate, was injected through a hole in the center of the tube so that the solution spread across the 180 µm-thick cellulose support to produce a thin aqueous film. After the support was loaded with the solution, clean gas was passed through the tube and the concentration of the released odorant measured by APCI-MS.

This system was used for several experiments, one of which is highlighted here. Specifically, measurement of the off-rate for butyl acetate was undertaken because the published value seemed low and did not align with the proposed mechanism of OBP in the mass transfer of odor compounds. Four isomers of butyl acetate were used as test VOCs to probe the affinity of the OBP binding site. The APCI-MS traces monitoring the release of *n*-butyl acetate from OBP solutions of different molarities were normalized and are shown in Figure 1. The cellulose support system allowed monitoring of the release of VOCs from an OBP solution, but it was so fast that accurate numerical values could not be assigned to the off-rate (9). This limitation and the desire to align the *in vitro* system more closely with the conditions and dimensions found *in vivo*, led to the design of a new system based on capillary tubes. This design replicated some but not all of the *in vivo* conditions.

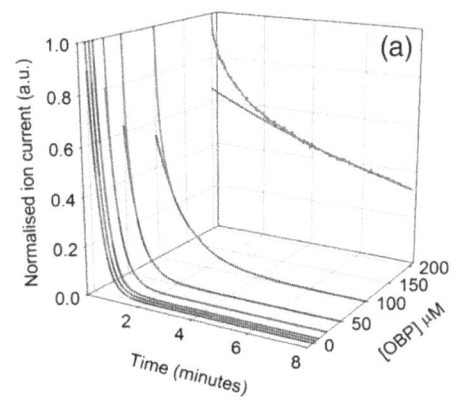

Figure 1. APCI-MS traces from the off-rate experiment for n-butyl acetate. Reproduced with permission from reference (9). Copyright 2010 Elsevier.

Binding and Release Measurements

As an alternative to the cellulose support, and to mimic more closely the conditions *in vivo*, fused silica capillary columns were internally coated with a thin film of rat OBP-3. This was achieved using GC capillary coating techniques, which are known to produce thin film coatings in the μm range, similar to the thickness of the OBP layer *in vivo* (*12*). The coverage and thickness of the OBP film in test capillaries was assessed using a colored solution of OBP and then by cutting the capillary into multiple short lengths and studying the thickness and integrity of the OBP layer using microscopy. With optimized coating parameters in place, a consistent coverage and film thickness could be established. The surface area and moles of OBP in the capillary could be calculated from the column and film dimensions.

Gas-tight syringes, mounted on syringe pumps, and containing solutions of VOCs, delivered either a constant gaseous stream of a VOC at known concentrations or a pulsed delivery by using a second syringe pump and a switching valve Figure 2. The second syringe could be used as a source of clean air or to deliver a second gaseous stream of a VOC when competition experiments were carried out.

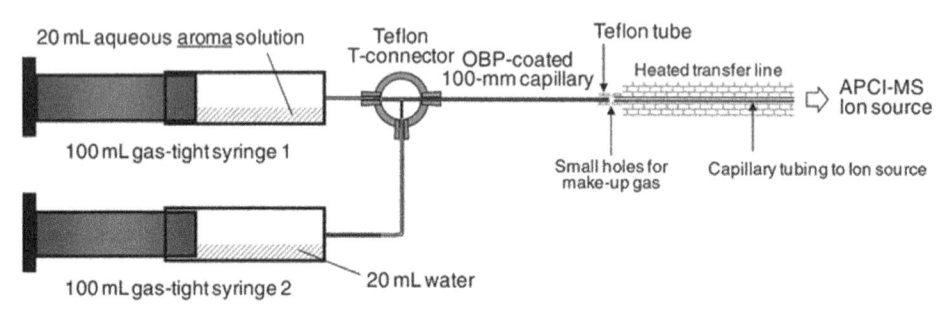

Figure 2. Schematic of the in vitro OBP system using a coated capillary column and gas-tight syringes for odorant delivery. Reproduced with permission from reference (12). Copyright 2010 Springer.

Table 2 lists some key factors that are found *in vivo* and compares these values with those that were achieved in the *in vitro* system. As with many experimental designs, there are some limitations and neither the flow rate nor the Reynolds number could be matched due to the dimensions of the capillary. For the APCI-MS requirements, compounds were chosen with different physicochemical properties but also chosen so their ions were differentiated to avoid the problem of matching ions to compounds (see Chapter 2).

APCI-MS monitored gas from the exit of the capillary and therefore measured the net binding/release of individual VOCs. Initially, the time taken to saturate the OBP in the capillary was established by waiting for "breakthrough" of the odorant to occur (as monitored by APCI-MS). Figure 3 shows the APCI-MS traces obtained for an OBP-coated capillary (trace a), a bovine serum albumin (BSA)-coated capillary (trace b) and a bare capillary (trace c) when isobutylthiazole (IBT) was the test odorant and introduced at time zero, then replaced with clean air at 15 min. The breakthrough time was significantly longer for the OBP-coated capillary, compared to a bare capillary or one containing BSA solution (*12*), showing that OBP bound the VOC to a much higher extent than BSA and there was no binding to the capillary itself. The dynamic binding and release behavior of other VOCs with different physicochemical properties has been published (*17, 18*).

Table 2. Conditions in the peri-receptor region relevant to odor uptake and the corresponding conditions in the capillary *in vitro* system

Factor	In vivo	In vitro
Thickness of mucosal layer (μm)	1-5 (*13, 14*)	1.3
Surface area (cm^2)	1-5 (*15*)	1.7
OBP concentration (mM)	1 (*3*)	0.1-1.25
Reynolds number dimensionless	610-980 (*16*)	3-51
Gas linear velocity (cm/s)	53-147 (*16*)	8-151
Flow rate (mL/s)	125-200 (*16*)	0.017-0.33

Numbers in parentheses identify the source of the information

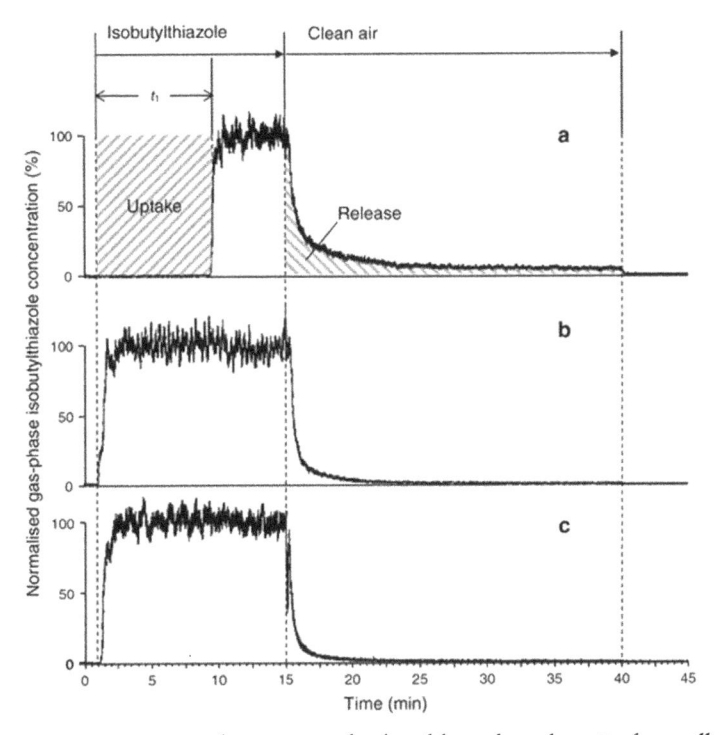

Figure 3. APCI-MS trace monitoring the presence of isobutylthiazole at the exit of a capillary loaded with a thin film of rat OBP (trace a), a thin film of bovine serum albumin (trace b) and a bare capillary (trace c). APCI-MS is recording the net binding/release of the volatile compound as a function of time. Reproduced with permission from reference (12). Copyright 2010 Springer.

The first learning from Figure 3 is that OBP completely binds the incoming IBT for nearly 8 min, whereas the BSA and the bare column showed no IBT binding. Of more interest, however, is the data processing needed to interpret the APCI-MS trace data into binding and off-rate values. To calculate the number of moles delivered to the OBP and the number of moles released, it is the areas marked "uptake" and "release" in Figure 3 that are the numerical values that need to be extracted.

According to its dissociation constant, IBT has a high affinity for OBP, and Figure 3 clearly shows that binding is *the* major event and that release is a very long and slow process. Compounds that bind less strongly will be expected to release more quickly. Since OBP was coated at 1 mM and the volume of solution in the capillary was known, it is possible to calculate the number of moles of OBP in the system and compare that with the number of moles of IBT bound to determine the number of binding sites per mole of OBP. With a range of odorant concentrations, the stoichiometry of IBT binding was 0.98, suggesting a 1:1 binding ratio of odorant molecules to the OBP molecule, a value that agrees with studies on OBPs from other species (19) and acts as validation that the *in vitro* system produces data consistent with other techniques.

The capillary *in vitro* system described above was used to further examine the binding and release of odor compounds over short time periods (0-60 s). For these experiments, a pulse of IBT at a concentration representative of that found during food consumption (22 nM) was introduced to load the OBP. Because IBT was bound to the OBP, no IBT was detected by APCI-MS monitoring of the outflow (Figure 4; trace up to time zero). At time zero, different flow rates of clean air were passed through the capillary and IBT was observed to be "eluted" at different rates, as expected. Figure 4 shows that higher flow rates elute IBT in peaks with a higher signal-to-noise ratio than low flow rates. This *in vitro* experiment could be interpreted as the situation that occurs when humans sniff to detect an odor by accelerating the airflow through the nasal passages and thereby concentrating odors absorbed onto the nasal mucosal layers into a single sharp peak with high signal-to-noise ratio to aid detection of faint smells by the odor receptors.

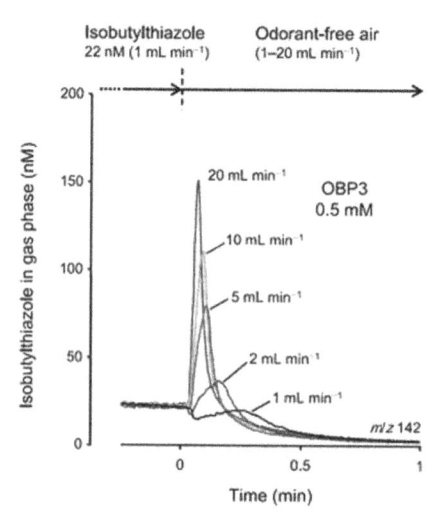

Figure 4. Release of OBP-bound isobutylthiazole (detected in APCI-MS at m/z 142) showing the effect of different air flow rates on the elution profiles. Reproduced with permission from reference (17). Copyright 2010 Oxford Academic.

The effect of competition between odorants was examined by choosing two compounds (2-isobutyl-3-methoxypyrazine (IBMP) and IBT), that had similar water solubility values but very different dissociation constants for OBP. Therefore, the rate of uptake into the aqueous layer should be similar for the two compounds but binding and release from OBP should be different.

The two compounds could be easily differentiated in the APCI-MS spectra by monitoring *m/z* 142 for IBT and *m/z* 167 for IBMP. Figure 5 shows the results of a preliminary experiment where a 3-min flow of IBMP (500 ppb) was passed over the OBP and then replaced by clean air (Figure 5A). For the first 3 min, the APCI-MS trace at *m/z* 167 flat-lined because all the IBMP being introduced was bound by the OBP and no IBMP reached the APCI-MS. When clean air replaced the IBMP flow, the OBP started to release the bound IBMP and the *m/z* 167 trace shows the release curve. In Figure 5B, after 3 min of IBMP introduction, the flow was replaced with a flow of IBT (500 ppb) and the release curve shows active displacement of IBMP compared to the air control and binding of IBT up to around 7.5 min when the OBP has become saturated with IBT and it appears in the outflow gas stream. The experiment was repeated using sequential exposure to IBMP and IBT at 1 min durations and over four cycles to mimic the tidal flow of breath, albeit at longer time intervals. Detailed results can be found in the literature (*17*).

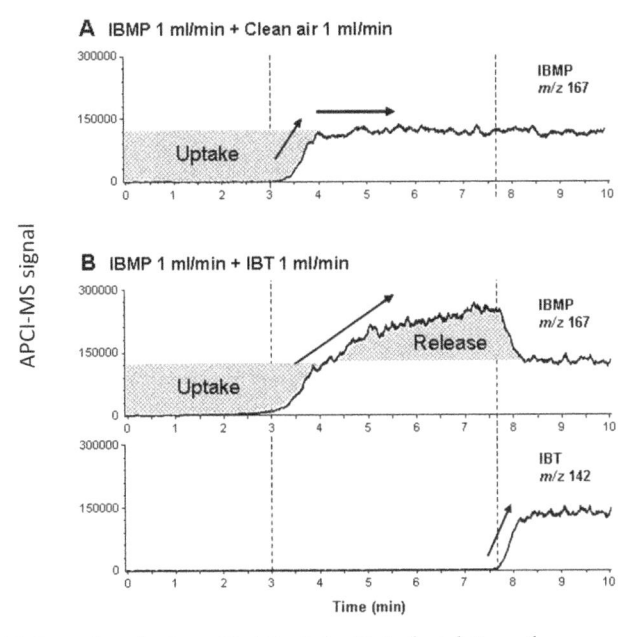

Figure 5. A: Uptake (0-3 min) and release (3-10 min) of 2-isobutyl-3-methoxypyrazine (IBMP) at a flow rate of 1 mL/min. B: Uptake and release of IBMP when clean air was replaced by isobutylthiazole (IBT), showing that IBT binding displaces IBMP faster and to a greater extent than air. The upper B trace monitors IBMP at m/z 167, the lower B trace monitors IBT at m/z 142. Dashed vertical lines show the onset of uptake/release over the three traces.

Conclusions

The ability to monitor the binding and release of selected odor compounds from OBP at concentrations typically found *in vivo* opens new opportunities to study the dynamics of odor binding in the peri-receptor region. The flexibility of APCI-MS with regard to sample flow rate, the humidity of the sample, and its ability to monitor ions at specific settings (cone voltages) to aid assignment of ions to compounds, makes it well-suited for studies on the micro scale. A complementary study on odor metabolism in the olfactory tissue using proton transfer reaction-mass spectrometry (PTR-MS) has also delivered a fascinating insight into the mechanisms that occur (*20*). These applications of real-time MS to study physiological processes extend the usefulness of direct MS and may also lead to their application in medicine, as OBP has been proposed as a natural treatment for some pathogenic yeast diseases, such as *Candida albicans*. The yeast cells proliferate in humans through the quorum-sensing mechanism that is driven by an organic compound, farnesol (see for example (*21*)), and OBP has been shown to effectively bind farnesol and therefore could contribute to a decrease in the spread of the fungus.

Acknowledgments

The contributions of Loic Briand, Tony Borysik, David Scott and Rob Linforth, as well as financial support from BBSRC, UK and Kao Corporation, Japan are acknowledged.

Abbreviations

APCI-MS	Atmospheric Pressure Chemical Ionization-Mass Spectrometry
BSA	Bovine Serum Albumin
DIMS	Direct Injection Mass Spectrometry
IBMP	IsoButyl Methoxy Pyrazine
IBT	IsoButyl Thiazole
MS	Mass Spectrometry
OBP	Odor Binding Protein
PTR-MS	Proton Transfer Reaction-Mass Spectrometry
SIFT	Selected Ion Flow Tube-Mass Spectrometry
VOC	Volatile Organic Compound

References

1. de Souza, F. M. S.; Antunes, G. Biophysics of olfaction. *Reports on Progress in Physics* **2007**, *70* (3), 451–491.
2. Wang, B.; Cao, S.; Liu, W.; Wang, G. R., Effect of OBPs on the response of olfactory receptors. In *Odorant Binding and Chemosensory Proteins*; Pelosi, P., Knoll, W., Eds.; 2020; Vol. 642, pp 279–300.
3. Pelosi, P. Perireceptor events in olfaction. *J. Neurobiol.* **1996**, *30* (1), 3–19.
4. Laing, D. G.; Macleod, P. Reaction Time For the Recognition of Odor Quality. *Chem. Senses* **1992**, *17* (3), 337–346.
5. Pevsner, J.; Snyder, S. H. Odorant-Binding Protein - Odorant Transport Function in the Vertebrate Nasal Epithelium. *Chem. Senses* **1990**, *15*, 217–222.
6. Briand, L.; Nespoulous, C.; Perez, V.; Remy, J. J.; Huet, J. C.; Pernollet, J. C. Ligand-binding properties and structural characterization of a novel rat odorant-binding protein variant. *Eur. J. Biochem.* **2000**, *267* (10), 3079–3089.
7. Charlier, L.; Nespoulous, C.; Fiorucci, S.; Antonczaka, S.; Golebiowski, J. Binding free energy prediction in strongly hydrophobic biomolecular systems. *Physical Chemistry Chemical Physics* **2007**, *9* (43), 5761–5771.
8. Golebiowski, J.; Antonczak, S.; Fiorucci, S.; Cabrol-Bass, D. Mechanistic events underlying odorant binding protein chemoreception. *Proteins-Structure Function and Bioinformatics* **2007**, *67* (2), 448–458.
9. Borysik, A. J.; Briand, L.; Taylor, A. J.; Scott, D. J. Rapid Odorant Release in Mammalian Odour Binding Proteins Facilitates Their Temporal Coupling to Odorant Signals. *J. Mol. Biol.* **2010**, *404*, 372–380.
10. Taylor, A. J.; Cook, D. J.; Scott, D. J. Role of Odor Binding Protein: comparing hypothetical mechanisms with experimental data. *Chemosensory Perception* **2008**, *1*, 153–162.

11. Lobel, D.; Strotmann, J.; Jacob, M.; Breer, H. Identification of a third rat odorant-binding protein (OBP3). *Chem. Senses* **2001**, *26* (6), 673–680.
12. Yabuki, M.; Portman, K. L.; Scott, D. J.; Briand, L.; Taylor, A. J. DyBOBS: A dynamic biomimetic assay for odorant binding to Odor Binding Protein. *Chemosensory Perception* **2010**, *3* (2), 108–117.
13. Menco, B. P. M. Qualitative and quantitative freeze-fracture studies on olfactory and nasal respiratory epithelial surfaces of frog, ox, rat, and dog .2. Cell apices, cilia, and microvilli. *Cell and Tissue Research* **1980**, *211* (1), 5–29.
14. Reese, T. S. Olfactory Cilia in Frog. *Journal of Cell Biology* **1965**, *25* (2P2), 209–230.
15. Guntherschulze, J. Studies about the regio olfactoria of the wild boar (Sus-scrofa) and of the domestic pig (Sus-scrofa-domestica). *Zoologischer Anzeiger* **1979**, *202* (3-4), 256–279.
16. Keyhani, K.; Scherer, P. W.; Mozell, M. M. Numerical simulation of airflow in the human nasal cavity. *Journal of Biomechanical Engineering-Transactions of the ASME* **1995**, *117* (4), 429–441.
17. Yabuki, M.; Scott, D. J.; Briand, L.; Taylor, A. J. Dynamics of Odorant Binding to Thin Aqueous Films of Rat-OBP3. *Chem. Senses* **2011**, *36* (7), 659–671.
18. Yabuki, M.; Scott, D. J.; Briand, L.; Taylor, A. J., Measuring dynamics of odorant binding to odorant binding protein under physiological conditions. In *Advances and Challenges in Flavor Chemistry and Biology*; Hoffman, T., Meyerhof, W., Schieberle, P., Eds.; Deutsche Forschungsanstalt fuer Lebensmittelchemie: Freising, 2011; pp 33–38.
19. Vincent, F.; Spinelli, S.; Ramoni, R.; Grolli, S.; Pelosi, P.; Cambillau, C.; Tegoni, M. Complexes of porcine odorant binding protein with odorant molecules belonging to different chemical classes. *J. Mol. Biol.* **2000**, *300* (1), 127–139.
20. Schoumacker, R.; Robert-Hazotte, A.; Heydel, J. M.; Faure, P.; Le Quere, J. L. Real-time monitoring of the metabolic capacity of ex vivo rat olfactory mucosa by proton transfer reaction mass spectrometry (PTR-MS). *Anal. Bioanal. Chem.* **2016**, *408* (6), 1539–1543.
21. Riboni, N.; Spadini, C.; Cabassi, C. S.; Bianchi, F.; Grolli, S.; Conti, V.; Ramoni, R.; Casoli, F.; Nasi, L.; Fernandez, C. D.; Luches, P.; Careri, M. OBP-functionalized/hybrid superparamagnetic nanoparticles for Candida albicans treatment. *RSC Advances* **2021**, *11* (19), 11256–11265.

Chapter 7

APCI-MS/MS—An Enhanced Tool for the Real-Time Evaluation of Volatile Isobaric Compounds

Ni Yang,[*] Clive Ford, and Ian Fisk

Food Flavour Group, Division of Food, Nutrition and Dietetics, University of Nottingham, Sutton Bonington Campus, LE12 5RD Loughborough, United Kingdom
[*]Email: ni.yang@nottingham.ac.uk

Atmospheric pressure chemical ionization-mass spectrometry (APCI-MS) can be used for the real-time analysis of volatile compounds both *in vitro* and *in vivo*. Recent developments of this technique, which are demonstrated herein, couple the APCI ion source and reaction chamber directly with tandem mass spectrometry (APCI-MS/MS). Nine aroma compounds were selected to demonstrate the benefit of the MS/MS approach using multiple reaction monitoring (MRM) mode compared to a single mass filtering step (single ion recording, SIR) in a static equilibrium system of corresponding aqueous solutions. The signal-to-noise ratio of this system operated in MRM mode was improved by a factor 1.5-25 compared with SIR mode. The MRM approach was also uniquely able to separate two isobaric aroma compounds (anisole and 2,5-dimethyl pyrazine), both *in vitro* (static headspace analysis) and *in vivo* (breath-by-breath). Overall, when APCI is coupled with tandem mass spectrometry, detection sensitivity was increased and isobaric compounds could be separated.

Introduction

Atmospheric pressure chemical ionization-mass spectrometry (APCI-MS) is well known for its ability to conduct non-destructive, direct and real-time assessment of gas-phase samples both *in vitro* and *in vivo* (*1*). Recent studies have shown how this technique can be applied to study the impact of capsaicin on aroma release *in vitro* and *in vivo* in a model system (*2*). Other recent *in vivo* examples of its use include a study on the effect of phenolic compounds from olive oil on aroma release and persistence (*3*), and *in vitro* examples include a recent study on the effect of oral physiology parameters on aroma release in the mouth (*4*). The data provided by APCI-MS analysis can also be used to predict sensory properties (e.g., of thickened model solutions (*5*)), product quality (e.g., of cheese maturation (*6*)) or to support flavor reformulation (e.g., of different fat levels (*7*)).

© 2021 American Chemical Society

Tandem mass spectrometry features a dual mass spectrometer (MS/MS) setup with a collision-induced dissociation (CID) chamber between the mass spectrometers that allows for structural identification. This is achieved by mass filtering specific precursor ions that subsequently undergo CID to form fragmentation patterns that can be unique to a molecule. This technique has been widely used in liquid chromatography (i.e., LC-MS/MS) and gas chromatography (i.e., GC-MS/MS). The benefit of coupling MS/MS with APCI-MS for real-time gas phase analysis (i.e., APCI-MS/MS) has been demonstrated in previous studies (8–10). The work reported in this chapter illustrates the step-by-step process of using APCI-MS/MS for studying volatile flavor release *in vivo* and *in vitro*.

Altering the operating mode of APCI-MS/MS changes the behavior of ions through the triple quadrupole. Operating in scan mode stabilizes the trajectory of all ions within a specified *m/z* range through the quadrupoles, producing a full spectrum within that range. Single ion recording (SIR) mode features a single mass filtering step at a specific *m/z*, whereby mass filtering is achieved either with a single quadrupole or with all quadrupoles in tandem. Product scan mode induces fragmentation in a specified precursor ion and generates a spectrum for all product ions. Precursor ion refers to the initial ion of interest, usually the adduct ion generated in the initial ionization step. Multiple reaction monitoring (MRM) mode with a highly selective detection protocol can monitor the precursor-product ion transitions. A schematic of APCI-MS/MS is presented in Figure 1.

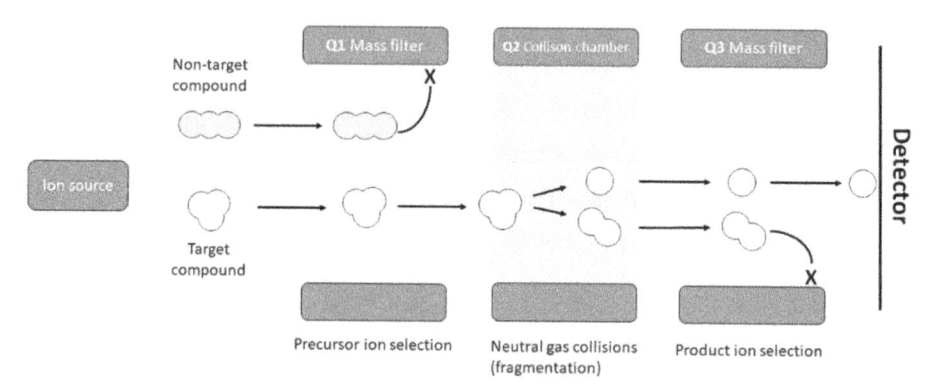

Figure 1. Schematic for a triple quadrupole mass spectrometer, including the first quadrupole (Q1), the second quadrupole (Q2), and the third quadrupole (Q3).

As is shown in Figure 1, the target compound enters the ion source and precursor ions are mass filtered at the first quadrupole (Q1). Fragmentation is carried out in the CID chamber at the second quadrupole (Q2) or hexapole situated at Q2. The collision cell is pressurized with an inert collision gas, such as argon, to fragment the precursor ion to characteristic product ions. The selected product ions can be mass filtered at the third quadrupole (Q3) and then transferred to the detector. Therefore, each quadrupole has its major function: Q1 is responsible for mass filtering for the precursor ion, Q2 induces fragmentation of precursor ions to product ions, and Q3 filters for the desired product ion.

One advantage of using APCI-MS/MS over conventional APCI-MS is its capability of enhancing selectivity and lowering the limit of detection (LOD) for target compounds, in particular in MRM mode. Another distinctive advantage of APCI-MS/MS is its potential to separate isobaric compounds, which is not possible in conventional APCI-MS although could be similarly achievable by replacing the quadrupole mass filter with a time-of-flight MS, as has been successfully demonstrated in proton transfer reaction time-of-flight mass spectrometry (PTR-TOF-MS) for isobaric flavor compounds (*11*).

Hence, the aims of the present study were twofold: the first objective was to validate if MRM mode exhibited higher selectivity and sensitivity over SIR mode; the second objective was to develop a solid approach of using APCI-MS/MS to separate two isobaric compounds both *in vitro* and *in vivo*.

Experimental Methods

Food-grade chemicals, including 2,5-dimethyl pyrazine (98%), 2-pentanone (≥98%), ethyl acetate (99%), butyl formate (97%), methyl butyrate (99%), ethyl butyrate (99%), anisole (99%), 2-octanone (≥95%) and isoamyl acetate (≥95%) were obtained from Sigma Aldrich (Dorset, UK). These selected compounds had different physicochemical properties, with estimated hydrophobicity (Log P) and vapor pressure (VP) values calculated using EPISuite version 4.10 (US Environmental Protection Agency), as listed in Table 1.

Each aroma compound was diluted in pure water (Purite Ltd, Oxon, UK) in a 100 mL Duran screw-capped bottle that was then placed on a roller mixer (Tube Roller Spiramix 10, Thermo Scientific, Loughborough, UK) to roll and shake for 1 h. The final concentration for each compound used for APCI-MS/MS analysis is listed in Table 1.

An initial full mass scan ranging from m/z 0-200 was performed in order to identify volatile adduct ions in preparation for analysis in SIR mode. Signal intensity was initially assessed for all volatiles at a cone voltage of 20 V, after which it was adjusted (10-50 V) for each compound to achieve the optimal voltage that elicited the highest signal. The procedure to optimize the detection of volatiles by modulating the cone voltage to increase the associated m/z signal (also known as decluttering potential) has been described in the literature (*12*).

The MRM analysis mode was used for the identification of appropriate precursor-to-product ion transitions using argon as the neutral collision gas. Using the optimal cone voltage for each compound, a range of collision energies (5-30 eV) was evaluated for each selected ion to determine the optimal setting. i.e., the energy eliciting the highest signal intensity (Table 1).

Static Headspace Aroma Analysis (*In Vitro*)

A static headspace sampling protocol was used to measure partitioning of volatiles in aqueous solutions at equilibrium using APCI-MS/MS (Waters TQD prototype, Manchester, UK). Standard solutions of 50 mL (three replicates) of each aroma compound were equilibrated for a minimum of 3 h at 25 °C in 100 mL Duran bottles before headspace analysis. Caps of the Duran bottles (100 mL) were modified with a sealable sampling port for connection of the APCI-MS/MS sampling line. The sampling rate was 10 mL/min for approximately 1 min through the transfer line that housed a silica GC column of 0.25 μm thickness and 0.25 mm inner diameter (Phenomenex, Macclesfield, UK). The transfer line temperature was set at 120 °C, the corona pin voltage was 4 kV operated in positive ion mode, and a dwell time of 0.1 s was used for all acquisitions.

For the measurement of the two isobaric compounds (anisole and 2,5-dimethyl pyrazine), individual solutions of each compound (2 ppm and 50 ppm, respectively) were used to establish the detection protocol for each compound in the MRM mode of APCI-MS/MS. The maximum ion intensity for each compound from each sample was recorded. Then, a mixture of the two compounds (50 mL, three replicates) was made with the same final concentration as in their separate solutions, and their partitioning at equilibrium was measured with APCI-MS/MS using the same settings as for the analyses of the individual solutions.

Table 1. Nine selected aroma compounds with different physicochemical properties and their respective concentrations used for APCI-MS/MS analysis with optimal cone voltage and collision energy, precursor ion, and product ion.

Compound	Log P[a]	VP (est. 25 °C)[b]	Concentration (ppm)	Cone voltage (V)	Collision energy (eV)	Precursor ion (m/z)	Product ion (m/z)
2,5-Dimethyl pyrazine	0.63	119	50	48	20	109	82
2-Pentanone	0.75	39.4	20	20	15	87	45
Ethyl acetate	0.86	98.3	1	20	20	89	61
Butyl formate	1.19	78.6	20	10	10	103	57
Methyl butyrate	1.36	33.2	5	29	10	102	71
Ethyl butyrate	1.85	14.6	1	25	10	117	89
Anisole	2.07	3.38	2	32	20	109	94
2-Octanone	2.22	1.87	1	29	15	129	44
Isoamyl acetate	2.26	5.67	1	15	10	131	70

[a] Log P is the logarithm of the partitioning coefficient of the molecule between octanol and water, which was estimated by the KOWWIN program (13). [b] VP is the estimated vapor pressure (mm Hg) of the molecule at 25°C, which was estimated by the mean VP of the Antoine and Grain methods (14).

Breath-by-Breath Analysis (*In Vivo*)

The *in vivo* release of the isobaric aroma compounds anisole and 2,5-dimethyl pyrazine was evaluated. Compared to the headspace samples, a higher concentration of each solution was required to obtain strong signals *in vivo*; thus, 50 ppm anisole and 150 ppm 2,5-dimethyl pyrazine (three replicates each) aqueous solutions were prepared separately with pure water (Purite Ltd, Oxon, UK). Then a mixture containing both compounds (three replicates) at the same final concentrations as their individual solutions was prepared. All samples (50 mL) were equilibrated at 70°C for 3 h in 120 mL Duran bottles and their headspace was extracted by a Luer lock gas-tight syringe and subsequently injected directly into the panelist's nasal cavity during inhalation.

Three healthy panelists (2 male, 1 female, 25-32 years old) were recruited from the University of Nottingham. Panelists were trained to maintain normal breathing through their nostrils directly into a sampling interface of the APCI-MS/MS, known as MS-Nose, during the injection period. The breath-by-breath intensities of the two compounds were monitored after the injection of the headspace containing the aroma compound(s) (individual or mixed solutions) at a sampling rate of 30 mL/min. Acetone (m/z 59), an endogenous compound present in breath, was continuously monitored in each panelist's breath as an analytical approach to follow the breathing pattern. A timed breathing protocol was applied, comprising the exhalation into the MS-nose after sample injection, then following a normal breathing pattern for 60 s for anisole or 90 s for 2,5-dimethyl pyrazine. Every panelist received each sample three times in a randomized order and had at least a 15 min break between each measurement. Before injection of the next sample, panelists were asked to breathe into the MS-Nose to ensure that no residue of the target compounds was present in their background breath.

Data Analysis

The maximum intensity (Imax) of the ion corresponding to each aroma compound was recorded for every study. The intensity of the noise signals was also recorded to calculate the signal-to-noise (S/N) ratio. Analysis of variance (ANOVA) was performed by the XLSTAT software (Addinsoft Inc, Paris, France). The level of significance was set at $p < 0.05$.

Aroma release curves from Imax at the timed point during *in vivo* analysis were imported into the line fitting Prism 7 software (Graphpad Software, Cambridge, UK). Decay rates reported here refer to the period taking place from Imax to the end of the experiment, and decay curves were normalized to the Imax of each repetition, thus all other values are expressed relative to 1. The decay rate can be characterized by the rate constant k (s^{-1}), which was calculated by a non-linear regression for the one-phase decay model (Prism 7). This one-phase decay model can be described by equation (1),

$$\frac{N}{N_0} = \exp^{-kt} \qquad (1)$$

where N is the level of aroma released at time t (s), N_0 is the aroma level released at t = 0, and k is the rate constant (s^{-1}).

Results and Discussion

Enhancing Signals and Sensitivity

It has been proposed that the MRM mode in APCI-MS/MS can achieve increased signal intensities and higher sensitivity than in SIR mode, corresponding to lower limits of detection. This was explored by calculating the S/N achieved in the detection of nine aroma compounds measured via static headspace analysis in aqueous solutions in either SIR or MRM mode. Specifically, S/N was determined by calculating the ratio between the peak height (Imax) and the depth of the background noise to give an indication of instrumental sensitivity. The data from these studies is presented in Figure 2.

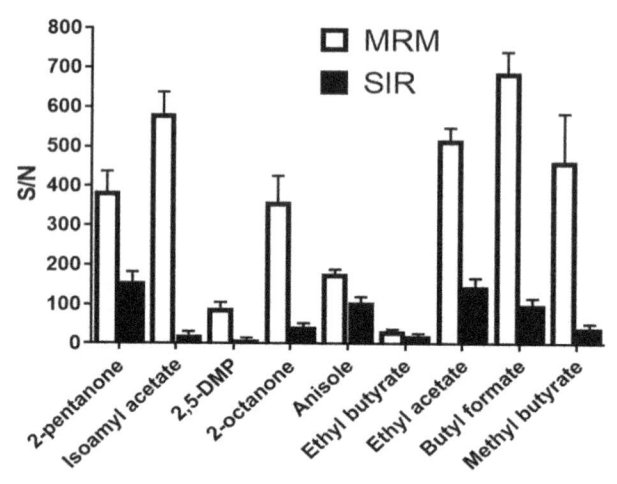

Figure 2. Calculated signal-to-noise (S/N) ratios for nine volatile compounds detected in either single ion recording (SIR) or multiple reaction monitoring (MRM) modes of APCI-MS/MS (n=3; error bars are ± standard deviation).

The S/N data indicate that MRM mode achieved significantly higher S/N ratios than SIR mode for all nine compounds ($p < 0.05$). Comparing the increased S/N ratios from SIR to MRM, anisole exhibited the lowest S/N improvement, with a factor 1.5 increase, and isoamyl acetate showed the best enhancement, with a factor 25 increase. This broad range is potentially attributable to the molecular structures of the individual aroma compounds and their ionization efficiency in APCI. Isoamyl acetate, as the ester formed from isoamyl alcohol and acetic acid, can be more easily ionized and form other product ions compared to anisole, which is an aromatic conjugated benzene ring attached with a methoxy group that is harder to transit to other product ions in MS/MS (8–10).

The enhanced sensitivity of an optimized MRM in tandem mass spectrometry has been reported in the literature (8–10). Generally, most aroma compounds in the present study exhibited an S/N increase of more than a factor 2 from SIR mode to MRM mode. The increased S/N in MRM mode indicated an improvement in the LOD and limit of quantification (LOQ), although further studies are required to establish absolute LOD and LOQ values for these and other isobaric compounds. Due to the significantly enhanced S/N in MRM mode, it is possible to detect those compounds when they are present at much lower concentrations, e.g., during breath-by-breath analysis.

Separating Isobaric Compounds in Headspace Analysis (*In Vitro*)

A highly selective method of using the MRM mode of APCI-MS/MS was required to separate the two selected isobaric compounds, anisole and 2,5-dimethyl pyrazine, both with a molecular weight of 108 g/mol. Upon soft ionization in SIR detection mode, anisole and 2,5-dimethyl pyrazine form adducts with a proton and appear at m/z 109. Under MRM mode, however, the fragmentation of each compound deviates. At the optimal conditions, anisole produced ions at m/z 77 and m/z 94, whereby the former was 58% more abundant than the latter, so m/z 77 was used as the main product ion. By comparison, 2,5-dimethyl pyrazine fragmented into two different product ions at m/z 41 and m/z 82, with the latter being 25% more abundant than the former, thus the ion at m/z 82 was chosen as the main product ion for 2,5-dimethyl pyrazine. Figure 3 demonstrates how different product ions were formed from precursor ions of anisole and 2,5-dimethyl pyrazine, respectively.

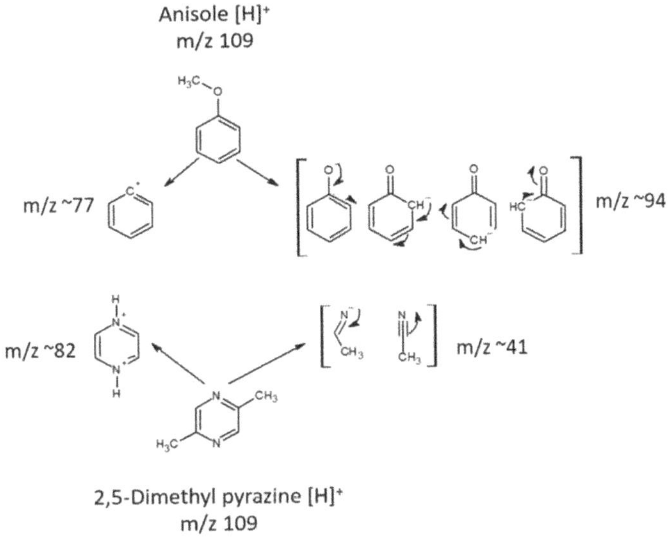

Figure 3. Precursor fragmentation of anisole and 2,5-dimethyl pyrazine

Ion stability and reactivity are important factors for proposing fragmentation pathways (*15*). For instance, charge retention fragmentation reactions informed the proposed fragmentation pathway of anisole (Figure 3). The charge stability of the delocalized π electron cloud prevents the migration of charge to the fragmentation reaction site and does not participate in the reaction (*16*). As proposed in Figure 3, the m/z of the major product ions of anisole (m/z 77 and m/z 94) can be attributed to the removal of the distal methyl group and the removal of the methyl ether group, respectively. Fragmentation of 2,5-dimethyl pyrazine produced m/z 82, which could potentially involve the remote hydrogen rearrangement reactions to protonate imine groups.

A mixture of anisole and 2,5-dimethyl pyrazine was used to validate if MRM mode with the selected ions could differentiate them. APCI-MS/MS was set to detect the transition of ions from m/z 109 to m/z 77 (anisole) and m/z 109 to m/z 82 (2,5-dimethyl pyrazine). The Imax values for each compound in individual solutions or as part of a mixture are summarized in Figure 4. There were no significant differences in signals generated for each compound between the separated solutions and the mixtures ($p > 0.05$). Therefore, monitoring specific molecular transitions using MRM mode in APCI-MS/MS can separate isobaric compounds, such as anisole and 2,5-dimethyl pyrazine. A wider range of isobaric compounds can be evaluated using the established APCI-MS/MS approach, such as the four pairs of compounds (cis-3-hexenol and 2,3-pentanedione, benzaldehyde and *m*-xylene, ethyl butanoate and 2-methylbutanol, and isobutyl isopentanoate and 1-hexanol) used in a previous PTR-TOF-MS study (*11*).

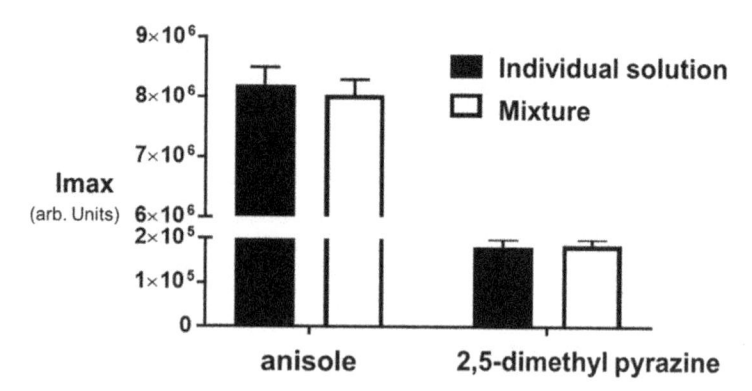

Figure 4. Imax values for anisole and 2,5-dimethyl pyrazine when assessed individually or as a mixture using static headspace analysis by APCI-MS/MS (n=3; error bars are ± standard deviation).

Separating Isobaric Compounds in Breath-by-Breath Analysis (*In Vivo*)

Since MRM mode demonstrated its success in separating the two selected isobaric compounds *in vitro*, the *in vivo* analysis was conducted to validate if these two compounds could be detected and separated after direct injection of their gas phase into the nostril. This direct injection method was designed to minimize any potential interactions of aroma compounds with the food matrix or saliva that are known to interfere with *in vivo* aroma release (*17, 18*). A similar method has been reported to study aroma persistence *in vivo* (*19*).

In-nose aroma intensity from breath-by-breath analysis for each compound, injected either individually or within a mixture, was averaged from three panelists and over three replicates (Figure 5). Similar patterns and concentrations of *in vivo* release were observed in the samples introduced individually or combined, irrespective of the compound. Therefore, with the appropriate MRM method, APCI-MS/MS is capable of separating isobaric compounds not only during headspace analysis, but also in breath-by-breath analysis.

As expected, the in-nose signals of both anisole and 2,5-dimethyl pyrazine decreased with time (as shown in Figure 5). However, the signals for anisole quickly returned to baseline within approximately 15 s, whereas this took over 80 s for 2,5-dimethyl pyrazine. This kinetic difference in aroma release can be described in terms of aroma persistence. These observations indicate that anisole is less persistent than 2,5-dimethyl pyrazine. Despite these volatiles being injected directly into the nasal cavity, some compounds might be absorbed in the nasal mucosa and epithelium, as proposed in other studies (*19, 20*). The physicochemical properties of the volatiles can be one of the main factors for their persistence in the breath. Linforth and Taylor (*20*) studied 41 compounds and illustrated that more polar compounds (low Log P) with high vapor pressure (higher VP) more readily partition into the mucous and persist longer in the breath. Comparing the estimated values of Log P and VP for 2,5-dimethyl pyrazine and anisole (see Table 1) it can be seen that 2,5-dimethyl pyrazine has a lower Log P and a higher VP than anisole, thus it can be expected to be more affected by the nasal mucus and, accordingly, have a longer persistence in breath than anisole. Indeed, this can be confirmed with the empirical data of the present study (see Figure 5).

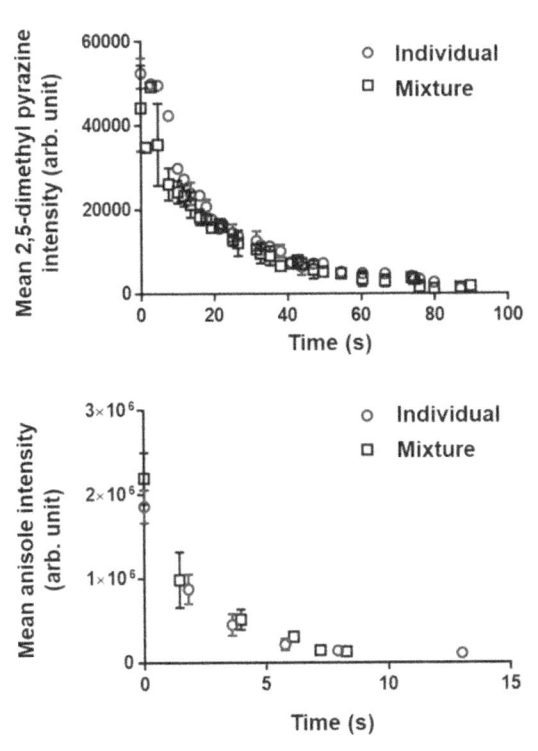

Figure 5. Mean in-nose aroma intensity of 2,5-dimethyl pyrazine and anisole after direct injection of each aroma into the nostril, either individually or as a mixture, during breath-by-breath analysis by APCI-MS/MS (three replicates per three panelists; error bars are ± standard deviation).

The persistence of aroma can be quantified by k (s^{-1}), which represents the decay constant and can be calculated from the associated curves of aroma intensity as a function of time. Analysis of the k (s^{-1}) values for each compound by ANOVA indicated that there was no significant difference between values derived from the individual solutions and those from the mixture ($p > 0.05$). A previous study similarly observed that inhaled compounds are retained in the airway mucosa and then released back to the breath flow (*19*). Further, it was found that compounds with different physicochemical properties had distinct decay rates: less hydrophilic but more volatile compounds (e.g., ethyl butyrate in their case, and anisole in the present case) exhibited a faster decay rate than the more hydrophilic but less volatile compound (2,5-dimethyl pyrazine in both studies). In addition, the impact of the physicochemical properties of different compounds on their persistence in breath was found to be much larger than the differences observed between individual panelists or in relation to different breathing protocols.

Different models have been used to predict aroma persistence effects in in-mouth flavor release (*20–22*). For example, a mechanistic model of flavor release was established based on interfacial mass transfer from liquids in the mouth, returning hydrophobicity and vapor pressure as driving factors (*20, 21*). Another study established different models on individual physicochemical properties, such as molecular weight, enthalpy of vaporization, VP, and Log P, that indicated that VP was the only parameter with statistical significance ($p < 0.01$) (*22*). However, these models were built on the retronasal aroma release after swallowing aqueous solutions, so a different theoretical model was established to predict the transport of inhaled volatile compounds in the nasal cavity (*23*). The APCI-MS/MS approach described in this chapter can be used to generate further data to validate these models.

Additionally, the isobaric compounds chosen in the present study are not isomers, i.e., they differ in their elemental composition, so they can also be separated in real-time by the higher resolving power of some mass spectrometers, such as TOF-based instruments (*11*). Future performance comparisons between TOF and MS/MS for aroma compound analysis would certainly be beneficial to choose the most suitable technology for a specific application. Another limitation of the present work is that only a single pair of isobaric compounds was investigated, thus future studies should include a larger suite of isobaric compound pairs and assess the capability of both techniques (TOF and MS/MS) to separate these compounds simultaneously in real-time.

Conclusions

Two advantages of APCI-MS/MS over conventional APCI-MS are showcased in this chapter. First, it was demonstrated how MRM detection protocols can be established for aroma headspace partitioning of nine selected aroma compounds from aqueous solutions. Using an optimal cone voltage and collision energy, the S/N ratio in MRM mode can be significantly enhanced compared to the SIR mode by more than a factor two for most compounds. This is because MRM has a dual-mass filtering procedure that allows the selection and detection of product ions specific to an individual compound. The degree of enhancement in the detection sensitivity varied between compounds, with a factor 1.5 increase in S/N for anisole and a factor 25 increase in S/N for isoamyl acetate. This difference is related to the molecular structures and the ionization efficiencies of the two compounds. The use of MRM mode in APCI-MS/MS with improved S/N *in vitro* also indicated that volatile compounds could be detected at lower concentrations, particularly in breath-by-breath analysis. Future studies to confirm the LOD and LOQ for a range of isobaric compounds would deliver insightful information for use of APCI-MS/MS in aroma release analysis.

Another notable advantage of APCI-MS/MS is its capability to separate isobaric compounds, as demonstrated in this chapter in both *in vitro* headspace analysis and *in vivo* nosespace analysis of anisole and 2,5-dimethyl pyrazine as representative compounds. The *in vivo* release results also illustrated the influence of the physicochemical properties of an aroma compound on their persistence in the nasal cavity, whereby the more hydrophilic and less volatile compound (2,5-dimethyl pyrazine) was observed to be more persistent in the nasal cavity than the less hydrophilic and more volatile compound (anisole).

This newly developed APCI-MS/MS technique, which offers enhanced sensitivity and extra capability for detecting isobaric compounds, represents a useful tool for future studies to provide additional data for modeling and predicting aroma release and persistence.

Acknowledgments

This work was supported by the Engineering and Physical Sciences Research Council (grant number: EP/M506588/1, 2014). The authors would like to thank the panelists for their time and support.

References

1. Taylor, A. J.; Linforth, R. S. T.; Harvey, B. A.; Blake, A. Atmospheric pressure chemical ionisation mass spectrometry for in vivo analysis of volatile flavour release. *Food Chem.* **2000**, *71* (3), 327–338.

2. Yang, N.; Galves, C.; Racioni Goncalves, A. C.; Chen, J.; Fisk, I. Impact of capsaicin on aroma release: in vitro and in vivo analysis. *Int. Food Res. J.* **2020**, *133*, 109197.
3. Genovese, A.; Caporaso, N.; di Bari, V.; Yang, N.; Fisk, I. Effect of olive oil phenolic compounds on the aroma release and persistence from O/W emulsion analysed in vivo by APCI-MS. *Int. Food Res. J.* **2019**, *126*, 8.
4. Tarrega, A.; Yven, C.; Semon, E.; Mielle, P.; Salles, C. Effect of Oral Physiology Parameters on In-Mouth Aroma Compound Release Using Lipoprotein Matrices: An In Vitro Approach. *Foods* **2019**, *8* (3), 15.
5. He, Q.; Hort, J.; Wolf, B. Predicting sensory perceptions of thickened solutions based on rheological analysis. *Food Hydrocoll.* **2016**, *61*, 221–232.
6. Gan, H. H.; Yan, B. N.; Linforth, R. S. T.; Fisk, I. D. Development and validation of an APCI-MS/GC-MS approach for the classification and prediction of Cheddar cheese maturity. *Food Chem.* **2016**, *190*, 442–447.
7. Yang, N.; Linforth, R. S. T.; Walsh, S.; Brown, K.; Hort, J.; Taylor, A. J. Feasibility of reformulating flavours between food products using in vivo aroma comparisons. *Flavour Frag J.* **2011**, *26* (2), 107–115.
8. Jublot, L.; Linforth, R. S. T.; Taylor, A. J. Direct atmospheric pressure chemical ionisation ion trap mass spectrometry for aroma analysis: Speed, sensitivity and resolution of isobaric compounds. *Int. J. Mass Spectrom.* **2005**, *243*, 269–277.
9. Le Quéré, J. L.; Gierczynski, I.; Sémon, E. An atmospheric pressure chemical ionization - ion-trap mass spectrometer for the on-line analysis of volatile compounds in foods: a tool for linking aroma release to aroma perception. *J. Mass Spectrom.* **2014**, *49*, 918–928.
10. Hatakeyama, J.; Taylor, A. J. Optimization of atmospheric pressure chemical ionization triple quadropole mass spectrometry (MS Nose 2) for the rapid measurement of aroma release in vivo. *Flavour Fragr. J.* **2019**, *34*, 307–315.
11. Zardin, E.; Tyapkova, O.; Buettner, A.; Beauchamp, J. Performance assessment of proton-transfer-reaction time-of-flight mass spectrometry (PTR-TOF-MS) for analysis of isobaric compounds in food-flavour applications. *LWT−Food Sci. Technol.* **2014**, *56* (1), 153–160.
12. Avison, S. J. Real-Time Flavor Analysis: Optimization of a Proton-Transfer-Mass Spectrometer and Comparison with an Atmospheric Pressure Chemical Ionization Mass Spectrometer with an MS-Nose Interface. *J. Agr. Food Chem.* **2013**, *61* (9), 2070–2076.
13. Linforth, R.; Cabannes, M.; Hewson, L.; Yang, N.; Taylor, A. Effect of Fat Content on Flavor Delivery during Consumption: an in Vivo Model. *J. Agric. Food Chem.* **2010**, *58*, 6905–6911.
14. Thomson, G. W. The Antoine Equation for Vapor-pressure Data. *Chem. Rev.* **1946**, *38*, 1–39.
15. Demarque, D. P.; Crotti, A. E. M.; Vessecchi, R.; Lopes, J. L. C.; Lopes, N. P. Fragmentation reactions using electrospray ionization mass spectrometry: an important tool for the structural elucidation and characterization of synthetic and natural products. *Nat. Prod. Rep.* **2016**, *33* (3), 432–455.
16. Tomer, K. B.; Gross, M. L. Fast atom bombardment and tandem mass spectrometry for structure determination: Remote site fragmentation of steroid conjugates and bile salts. *Biomed. Mass Spectrom.* **1988**, *15* (2), 89–98.
17. Siefarth, C.; Tyapkova, O.; Beauchamp, J.; Schweiggert, U.; Buettner, A.; Bader, S. Mixture design approach as a tool to study in vitro flavor release and viscosity interactions in sugar-free polyol and bulking agent solutions. *Int. Food Res. J.* **2011**, *44* (10), 3202–3211.

18. Ployon, S.; Morzel, M.; Canon, F. The role of saliva in aroma release and perception. *Food Chem.* **2017**, *226*, 212–220.
19. Sánchez-López, J. A.; Ziere, A.; Martins, S. I. F. S.; Zimmermann, R.; Yeretzian, C. Persistence of aroma volatiles in the oral and nasal cavities: real-time monitoring of decay rate in air exhaled through the nose and mouth. *J. Breath Res.* **2016**, *10* (3), 036005.
20. Linforth, R.; Taylor, A. J. Persistence of Volatile Compounds in the Breath after Their Consumption in Aqueous Solutions. *J. Agr. Food Chem.* **2000**, *48* (11), 5419–5423.
21. Wright, K. M.; Hills, B. P.; Hollowood, T. A.; Linforth, R. S. T.; Taylor, A. J. Persistence effects in flavour release from liquids in the mouth. *Int. J. Food Sci. Technol.* **2003**, *38* (3), 343–350.
22. Buffo, R. A.; Rapp, J. A.; Krick, T.; Reineccius, G. A. Persistence of aroma compounds in human breath after consuming an aqueous model aroma mixture. *Food Chem.* **2005**, *89* (1), 103–108.
23. Keyhani, K.; Scherer, P. W.; Mozell, M. M. A Numerical Model of Nasal Odorant Transport for the Analysis of Human Olfaction. *J. Theor. Biol.* **1997**, *186* (3), 279–301.

Chapter 8

From Mold Worms to Fake Honey: Using SIFT-MS to Improve Food Quality

Sheryl Barringer*

Department of Food Science and Technology, The Ohio State University, 2015 Fyffe Road, Columbus, Ohio 43210, United States
*Email: barringer.11@osu.edu

Selected ion flow tube-mass spectrometry (SIFT-MS) measures volatile organic compounds (VOCs) in the headspace or mouthspace of a sample. Rapidly and nondestructively measuring these aroma compounds in food can be useful in a number of different ways, including determining the causes of off-odor, understanding consumer acceptance of food, detecting adulteration, or identifying the original source of a food item. Several examples of using SIFT-MS in food quality are described in this chapter. Fruit juice sometimes suffers from off-odors associated with the growth of *Alicyclobacillus acidoterrestris*. SIFT-MS was used to demonstrate that vanillin is the substrate and guaiacol the off-odor, and to explore the effects of storage temperature and pH on microbial growth and off-odor production. SIFT-MS was used to analyze cheeses produced in different factories to understand the effects of processing steps on the biochemical pathways that result in specific volatiles that change the sensory perception of the cheese. The technique was also used to detect the adulteration of honey with corn syrup and the adulteration of olive oil with other types of oils. Finally, SIFT-MS has been employed to classify or identify the origin of products including coffee, honey, and garlic.

Determination of the Cause of Off-Odor in Fruit Juice

Shelf-stable fruit juices are a popular consumer choice because of their convenience. They are heat-treated to destroy most of the bacteria present, and the low pH of the juice provides an additional barrier to prevent pathogen growth. Nevertheless, there are some spoilage organisms that can survive the heat treatment and grow at low pH. Growth of these microorganisms produces unacceptable defects, including "mold worms" and off-odors. *Alicyclobacillus acidoterrestris* is a thermophilic spoilage organism that can survive pasteurization and has been found in juices with off-odors. Selected ion flow tube-mass spectrometry (SIFT-MS) was used to understand the conditions that cause the off-odors produced in fruit juices.

© 2021 American Chemical Society

Apple juice and other growth media were inoculated with *Alicyclobacillus acidoterrestris,* and the resulting volatiles in the headspace were measured (*1*). Guaiacol has a smoky, medicinal odor and was identified as the key volatile producing the off-odor. Due to the similarity of its structure, vanillin was a suspected substrate for its formation (Figure 1).

Figure 1. Proposed conversion of vanillin into guaiacol.

Inoculation with vanillin and *A. acidoterrestris* resulted in a linear correlation ($R^2 = 0.998$) between vanillin concentration and guaiacol release in the headspace of the growth medium, as measured by SIFT-MS (Figure 2). Vanillin is present naturally in fruit juice, thus guaiacol is produced even when additional vanillin is not added to the sample. The clear linear relationship strongly suggests that vanillin is the substrate that the microorganism is using to produce the off-odor.

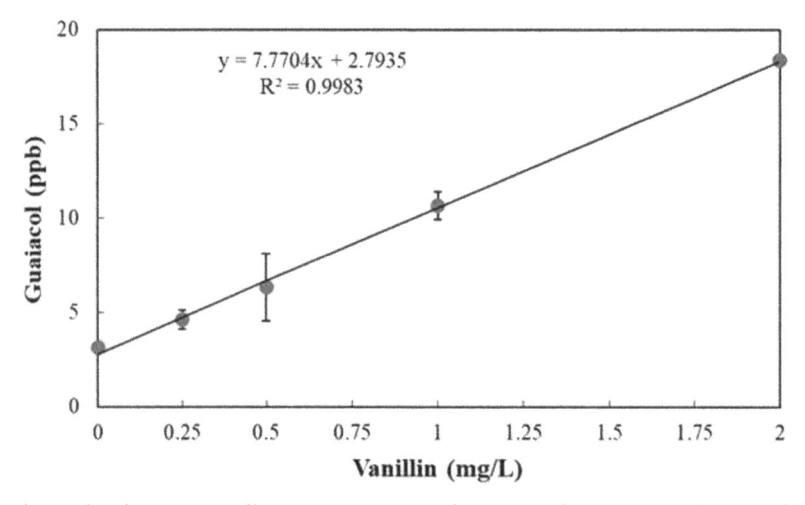

Figure 2. Relationship between vanillin concentration and amount of guaiacol produced and released into the headspace of the growth medium.

Since the microorganism *A. acidoterrestris* can survive pasteurization, it may be present as a contaminant in fruit juices. Storage temperature and pH were tested to explore the conditions under which the microorganism produces guaiacol, in order to stop its production. When apple juice was spiked with vanillin and *A. acidoterrestris* and stored at different temperatures, microbial growth correlated strongly with guaiacol production (Figure 3). At storage temperatures of 20 °C and 25 °C, microbial counts decreased and guaiacol production was prevented. At 37 °C and 45 °C, however, storage temperature affected the length of time for *A. acidoterrestris* to reach its stationary phase, as expected, but it did not significantly affect the final population size. The concentration of guaiacol in the headspace increased as microbial counts increased, then plateaued as microbial growth stagnated. The final concentration of guaiacol produced was affected by microbial count and was prevented at 25 °C and below. *A. acidoterrestris* is a thermophilic bacterium and its optimum growth temperature ranges from 42 °C to 60 °C (2). Thus, maintaining storage temperatures at or below room temperature will prevent formation of this off-odor.

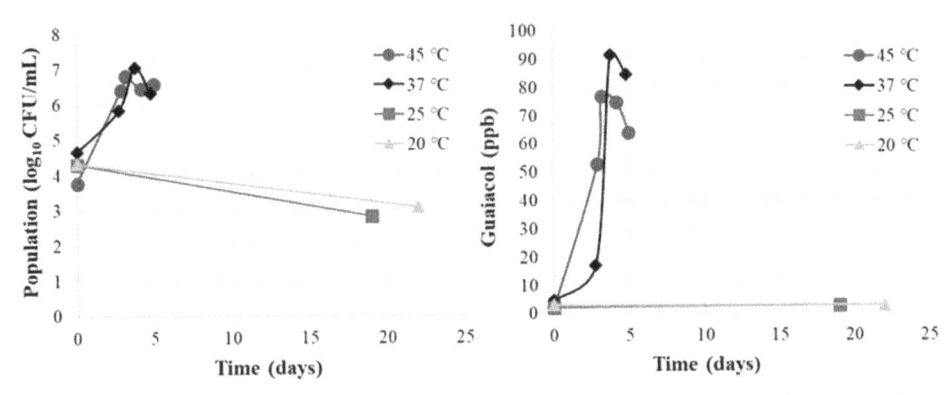

Figure 3. Growth of A. acidoterrestris (left) and guaiacol production (right) in apple juice in relation to storage temperature.

The pH of the juice was also found to affect both microbial growth and the subsequent guaiacol production. A pH close to neutral (6.7) inhibited the growth of *A. acidoterrestris* and thus no guaiacol was produced (Figure 4). Low pH (2.7) delayed bacterial growth and guaiacol production was delayed similarly. Bacterial growth in the exponential phase was the fastest at the pH values close to apple juice, pH 3.7 and 4.7, with corresponding fast guaiacol production. At pH 2.7, 3.7 and 4.7, both the final microbial counts and maximum guaiacol concentrations were similar.

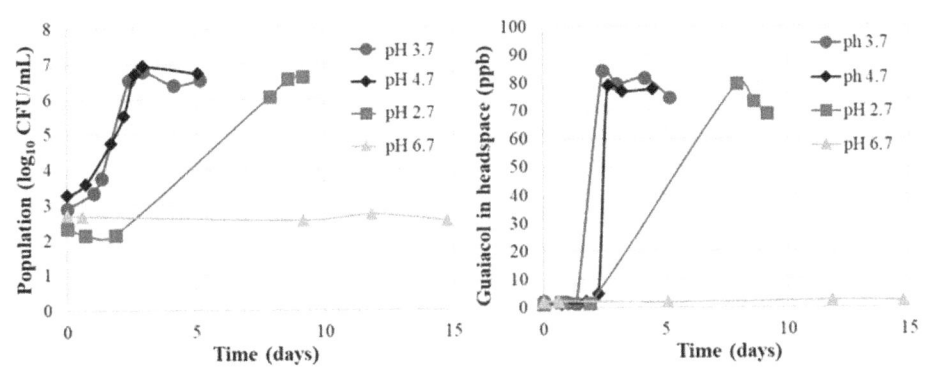

Figure 4. Growth (left) and guaiacol production (right) by A. acidoterrestris in apple juice during storage at different pH conditions.

Thus, SIFT-MS was a useful tool in understanding the cause of off-odors in fruit juice. The spoilage organism *A. acidoterrestris* produces guaiacol in fruit juices from vanillin that is naturally present in the juice. Volatile production can be stopped by keeping the juice at room or refrigeration temperatures, but the pH would have to be adjusted to an unacceptably high level before volatile production is stopped entirely.

Understanding Consumer Acceptance of Swiss Cheese

Consumers expect a consistent flavor of cheese produced by their preferred producer, so the company must ensure that it tastes the same, irrespective of the factory in which it was produced. If flavor varies by factory location, this is a problem because a company's reputation rests on producing a consistent product. By measuring the flavor volatiles in Swiss cheese, combined with principal component analysis (PCA), SIFT-MS can be used to determine if flavor varies based on factory, identify the biochemical causes of flavor differences, and predict sensory perception of those differences (3).

Using the concentration of key flavor volatiles present in Swiss cheese, statistical analysis was able to differentiate the cheeses produced in five different factories, by location (Figure 5). The next step was to determine what biochemical pathways produced the volatiles, and from there identify the processing steps that produced the differences in volatile levels. Correlations between the volatile compounds helped to identify the biochemical pathways and processing steps (Figure 5). For instance, there is a strong negative correlation between 3-methylbutanoic acid and 3-methylbutanal. During the ripening step, 3-methylbutanal oxidizes into 3-methylbutanoic acid, hence the negative correlation. Thus, the ratio of these two volatiles, which can be easily and nondestructively measured by SIFT-MS, is proposed as a fast way to determine the extent of ripening. When the cheese has not been ripened long enough, little flavor and aroma is present. When excessive oxidation occurs during the ripening step, this leads to the flavor defect oxidative rancidity. The ratio of 3-methylbutanoic acid to 3-methylbutanal can be used to predict when oxidative rancidity is imminent.

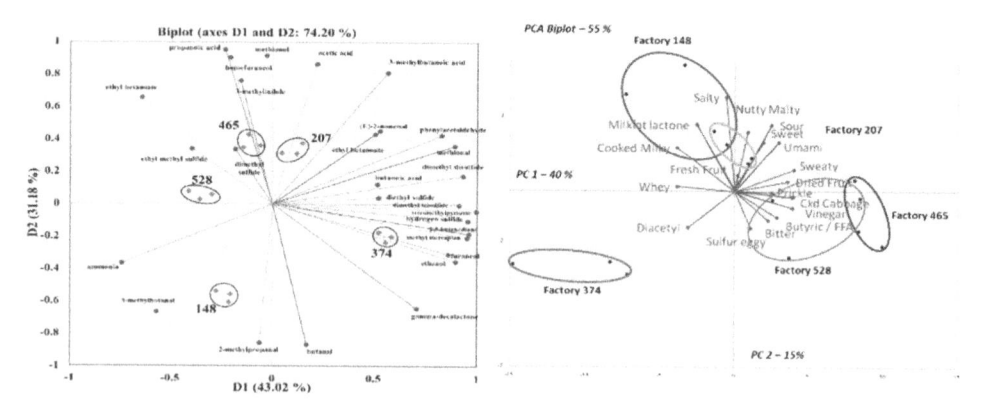

Figure 5. Principal component analysis showing differentiation of Swiss cheeses from different factories (148, 528, 465, 207, and 374) based on volatile compounds (left) and sensory descriptors (right).

On the other hand, the cause of the strong positive correlation between dimethyl disulfide, diethyl sulfide and butanoic acid is not immediately obvious. Dimethyl disulfide and diethyl sulfide are created from catabolism of the amino acid methionine, while butanoic acid is a free fatty acid generated by lipolysis of milk fat. Both processes occur during Swiss cheese ripening, creating many of the desirable flavors. The strong positive association of these volatiles indicates an association between proteolysis and lipolysis during the ripening of Swiss cheese that may be due to the lactic acid bacteria promoting both reactions, under favorable conditions.

A strong positive correlation was also observed between ethyl hexanoate and gamma-decalactone. Ethyl hexanoate is an ester with a fruity flavor produced from mono- and diglycerides. Gamma-decalactone is a lactone with a buttery flavor produced from triglycerides. The growth and metabolism of the bacteria in the cheese determines the activity of the lipolytic and esterolytic enzymes. The activity of the enzymes determines the extent of milk fat hydrolysis that occurs, to produce either mono- and diglycerides or leave the triacylglycerides intact. Thus, the conditions that promote the bacteria present in the cheese determines whether the cheese has high levels of fruity flavor from ethyl hexanoate, which can be a defect, or more buttery flavor from gamma-decalactone.

Correlations between volatile compounds and sensory attributes were used to better understand how the volatiles affect sensory perception (Figure 5). For instance, 2,3-butanedione correlates positively to the diacetyl sensory attribute. The correlation between 2,3-butanedione and the diacetyl sensory attribute is expected since 2,3-butanedione is another name for diacetyl and is the key volatile that produces the characteristic buttery aroma termed "diacetyl" by the panel. However, the fresh butter aroma of 2,3-butanedione is also inversely associated with the basic taste sensation elicited by sugars (sweet). In other words, the presence of 2,3-butanedione leads to a lower sweetness perception. It has previously been reported that high levels of 2,3-butanedione result in an unbalanced buttery flavor that lacks sweetness in butter.

Similarly, ethyl hexanoate is positively correlated with the sensory attributes dried fruit and sweet, but also with bitter. Ethyl hexanoate is perceived as fruity, thus its correlations with the sensory descriptors dried fruit and sweet are expected. High concentrations of ethyl hexanoate have been shown to produce fruity flavor defects in cheddar cheese, which may manifest as bitter in Swiss cheese. Gamma-decalactone positively correlated with cooked milky and diacetyl flavor attributes, while 2-methylpropanal and 3-methylbutanal have a strong positive correlation with the flavor attributes nutty-malty and salty.

Thus, SIFT-MS was used to determine the causes of flavor variability among Swiss cheeses produced in different factories. Starting at the end, with the sensory differences between cheeses from different factories, the company can work backwards through the volatiles, biochemical pathways and processing steps to determine the key conditions that must be kept consistent across factories in order to produce cheese with the same flavor. Specific sensory differences can be used to identify where variations in the processing occurred.

Adulteration

The National Center for Food Protection and Defense estimates that 10% of food in the grocery store is adulterated. The majority of the adulteration consists of intentionally adding a cheaper ingredient to the product in order to increase profitability. Honey and olive oil are estimated to constitute 7% and 14% of adulterated foods, respectively (4). SIFT-MS can be used to detect adulterated honey and olive oil by comparing the volatile pattern of authentic product to adulterated product (5–7). While there are no unique volatiles that are different, the concentrations and ratios of volatiles, or volatile patterns, can be differentiated. Honey is frequently adulterated with high fructose corn syrup because of the difficulty in detecting the difference based on the chemical composition of the sugars. However, statistical analysis of the volatile profile in the headspace of honey, determined by SIFT-MS, creates a clear separation of pure honey and adulterated honey in the PCA biplot (Figure 6). Similarly, extra virgin olive oil is frequently adulterated with lower grades of olive oil or oil from different sources. Again, statistical analysis of the volatile profile of olive oil can be used to separate olive oil from adulterated olive oil (Figure 6). With the huge economic incentive to adulterate food, companies struggle to find ways to test the authenticity of ingredients from suppliers, while regulatory agencies are also refining methodology to catch adulteration at the consumer level. SIFT-MS is a rapid, non-destructive way to test the authenticity of a product that is used by both companies and government agencies.

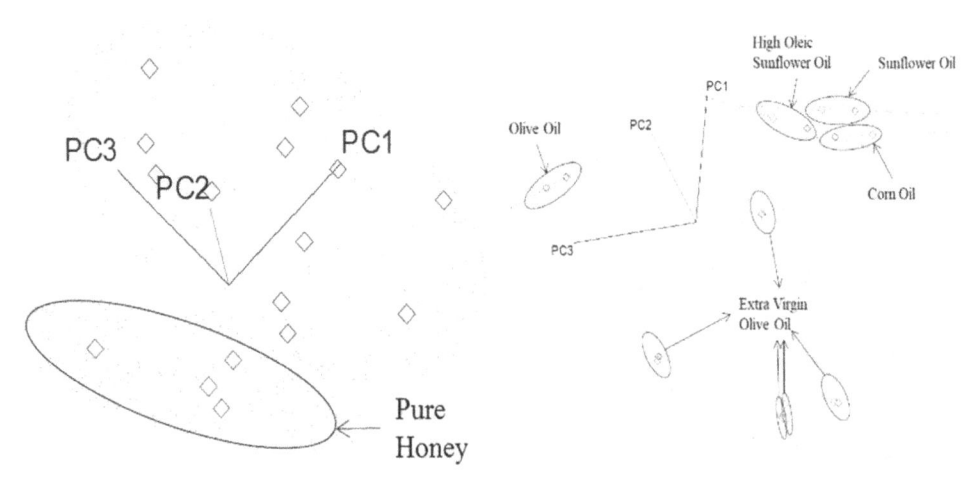

Figure 6. Principal component analysis showing differentiation of pure honey from honey adulterated with corn syrup (left), and pure olive oil from other oils (right), as based on volatile compounds in the headspace of the respective samples.

Classification

Identification of the exact source of a raw material is frequently important in food production. For a supplier, substituting one variety of coffee for another variety is one, often illegal, way to increase profit. Use of SIFT-MS for classification by source is a valuable way to maintain quality, and avoid regulatory problems. The volatile profile can be used to distinguish the origin of coffee beans or the exact variety of garlic (Figure 7) (8). In honey, the volatile profile can be used to distinguish honey by floral source, which is important in markets where honey from different floral sources can be sold at different prices (9, 10).

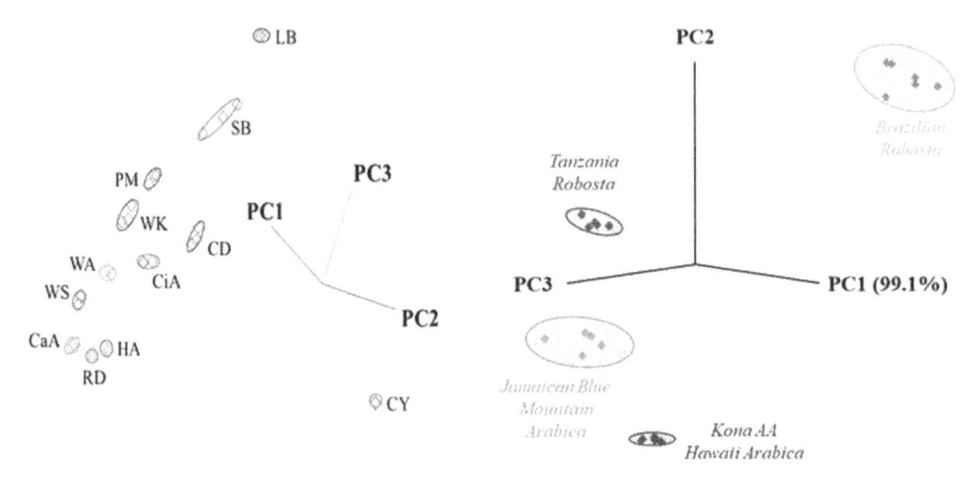

Figure 7. Principal component analysis showing differentiation of varieties of garlic (left) and coffee (right).

Conclusions

Selected ion flow tube-mass spectrometry is a rapid, non-destructive method for measuring volatiles in foods. Combined with statistical analysis of the volatile profile, and an understanding of the biochemical source and sensory impact of those volatiles, it becomes a powerful tool. SIFT-MS is used in production and quality control to maintain consistency. It can be used for research and development, to determine the cause of an off-odor that appears, or identify conditions to produce a product that better meets consumer expectations. Regulatory agencies find SIFT-MS a valuable tool for detecting adulteration and authenticating the source of food items. Similarly, it is used outside the food industry to measure a wide range of volatile organic compounds.

References

1. Huang, E.; Hu, X.; Barringer, S. A.; Yousef, A. E. Factors affecting *Alicyclobacillus acidoterrestris* Growth and Guaiacol Production and Controlling Apple Juice Spoilage by Lauric Arginate and ε-Polylysine. *LWT - Food Science and Technology* **2020**, *119*, 108883.
2. Chang, S. S.; Kang, D. H. *Alicyclobacillus* spp. in the fruit juice industry: History, characteristics, and current isolation/detection procedures. *Critical Reviews in Microbiology* **2004**, *30*, 55–74.
3. Castada, H. Z.; Hanas, K.; Barringer, S. A. Swiss Cheese Flavor Variability based on Correlations of Volatile Flavor Compounds, Descriptive Sensory Attributes, and Consumer Preference. *Foods* **2019**, *8*, 78.
4. Moore, J. C. Food Fraud: Public Health Threats and the Need for New Analytical Detection Approaches. In *Food Security: The Intersection of Sustainability, Safety and Defense*; 2011; pp 209–220.
5. Agila, A.; Barringer, S. Effect of Adulteration versus Storage on Volatiles in Unifloral Honeys from Different Floral Sources and Locations. *Journal of Food Science* **2013**, *78* (2), C184–C191.
6. Ozcan-Sinir, G.; Aykas-Cinkilic, D. P.; Barringer, S. A. Determination of Honey Adulteration Based on Volatile Profile using Selected Ion Flow Tube Mass Spectrometry (SIFT-MS) with Chemometrics. *Proceedings of the 1st International Congress on Analytical and Bioanalytical Chemistry (ICABC 2019)*, Antalya, Turkey, March 2019; pp 88–90.
7. Ozcan-Sinir, G. Detection of adulteration in extra virgin olive oil by selected ion flow tube mass spectrometry (SIFT-MS) and chemometrics. *Food Control* **2020**, *118*, 1–7.
8. Ozcan-Sinir, G.; Barringer, S. A. Variety differences in garlic sulfur compounds, by application of Selected Ion Flow Tube-Mass Spectrometry (SIFT-MS) with Chemometrics. *Turkish Journal of Agriculture and Forestry* **2020**, 408–416.
9. Agila, A.; Barringer, S. Application of Selected Ion Flow Tube Mass Spectrometry Coupled with Chemometrics to Study the Effect of Location and Botanical Origin on Volatile Profile of Unifloral American Honeys. *Journal of Food Science* **2012**, *77* (10), C1103–C1108.
10. Ozcan-Sinir, G.; Copur, O. U.; Barringer, S. A. Botanical and Geographical Origin of Turkish Honeys by Selected Ion Flow Tube-Mass Spectrometry and Chemometrics. *Journal of the Science of Food and Agriculture* **2020**, *100*, 2198–2207.

Chapter 9

Identifying Potential Volatile Spoilage Indicators in Shredded Carrot Using SIFT-MS

Lotta Kuuliala,[1,2,*] Nikita Jain,[1] Bernard De Baets,[2] and Frank Devlieghere[1]

[1]Research Unit Food Microbiology and Food Preservation (FMFP), Department of Food Technology, Safety and Health, Part of Food2Know, Faculty of Bioscience Engineering, Ghent University, Coupure links 653, B-9000 Ghent, Belgium
[2]Research Unit Knowledge-based Systems (KERMIT), Department of Data Analysis and Mathematical Modelling, Part of Food2Know, Faculty of Bioscience Engineering, Ghent University, Coupure links 653, B-9000 Ghent, Belgium
*Email: lotta.kuuliala@ugent.be

Minimally processed vegetables (MPV) are one of the most perishable food product types. The spoilage of these products is often due to microbial activity, which causes the accumulation of volatile organic compounds (VOCs) and leads to the generation of offensive off-odors and off-flavors in the packed product. Identification and quantification of VOCs that could be used as spoilage indicators can thus be considered a highly interesting strategy for improving MPV quality control. In the present study, shredded carrot (*Daucus carota* L.) was stored at 4 °C in a closed packaging system in order to characterize the volatilome associated with anaerobic spoilage processes. A multidisciplinary methodology based on selected ion flow tube-mass spectrometry (SIFT-MS) and probabilistic topic modeling was used for identifying potential spoilage indicators. The results of the study indicate that several VOCs contribute to the spoilage volatilome of carrots under the tested storage conditions. Even though multiple increasing patterns could be detected, the most abundant compounds typically remained stable over storage time. Overall, the developed methodology shows promising potential in extending the knowledge obtained via this preliminary study into new experimental settings.

Introduction

Plant-based foods form an essential part of a healthy and ecologically sustainable diet. One of the most popular groups within this sector comprises minimally processed vegetables (MPV), meaning vegetables that have undergone limited operations (e.g., washing, peeling, cutting and/or shredding) before packaging and refrigeration (*1*). This kind of processing not only promotes the diversity of plant-based products on the market, but also responds to increasing consumer demands for freshness, convenience and reduced preservative content.

One drawback of MPV, however, is that they tend to be more perishable when compared to their intact or highly processed (e.g., dried, frozen, fermented or canned) counterparts. Even though a minimal processing (MP) workflow typically consists of both value-adding and preservative operations, it also makes the products more susceptible to internal and external damage. Mechanical operations like cutting and shredding, for example, damage the plant tissues and expose larger surface areas to the surroundings, promoting both microbial growth and physiological deterioration (2). On the other hand, several studies have demonstrated cross-contamination risks associated with washing processes (3–5). Even though modified atmosphere packaging (MAP) can be used for extending the shelf-life of vegetables and fruit, finding and maintaining the optimal atmospheric conditions requires extensive effort. Because of these reasons, understanding the spoilage processes of MPV is of utmost importance for improving their quality control.

The formation of offensive off-odors is one of the most common reasons of food spoilage. This happens because of volatile organic compounds (VOCs): low-molecular-weight end-products of microbial and biochemical processes that accumulate in the package headspace over storage time and cause unpleasant sensory changes in the packed product. Consequently, identifying and quantifying spoilage-signifying VOCs has become a popular technique in food spoilage analysis. In particular, the development of real-time quantification techniques, such as selected ion flow tube-mass spectrometry (SIFT-MS) and proton transfer reaction-mass spectrometry (PTR-MS) has generated numerous insights into the volatilome of different food products, including vegetables, fruit and their processed derivatives (6–10).

Carrot (*Daucus carota* L.) is one of the most cultivated and commercially relevant vegetable products. Carrot is often sold in shredded or sliced form, typically as a part of mixed salads and ready-to-eat meals, but the shelf-life of these products is very limited. Several studies have assessed the microbiological spoilage of MP carrots under different packaging and storage conditions (11–14), yet analysis of the volatilome has been less frequently performed. Furthermore, even though solid-phase micro-extraction gas chromatography-mass spectrometry (SPME-GC-MS) and/or electronic nose (E-nose) sensory systems have been successfully applied for both minimally processed carrots (15, 16) and carrot-based products (17, 18), real-time mass spectrometry data is still scarcely available in the literature.

The aim of this study was to identify potential volatile spoilage indicators in shredded carrot stored at 4 °C for up to 9 days, using a multidisciplinary methodology combining real-time VOC monitoring (SIFT-MS), classical microbiological analysis, and a newly developed statistical method based on probabilistic topic modeling (19). Given the detrimental impact of anaerobic spoilage processes on the quality of MPV, shredded carrot was stored in a closed packaging system that allowed for the generation of anaerobic conditions over storage time.

Experimental Methods

Data Collection

Industrially shredded carrots packed in perforated plastic bags (500 g carrot per bag) were purchased from a local processing plant and delivered at the laboratory of Food Microbiology and

Food Preservation under chilled conditions. Immediately upon arrival (day 0), each bag was placed under atmospheric conditions in individual vacuum bags (PA/EVOH/PA/PE, O_2 transmission rate < 2.7 cm³/m²*24 h*bar at 23 °C and 0 % RH), heat-sealed and stored at 4.0 ± 0.3 °C for up to 9 days. At regular intervals (day 0, 2, 3, 5, 7 and 9), three randomly selected packages were analyzed for headspace gases, VOCs and microbial growth.

Headspace gas composition (% v/v O_2/CO_2) was determined with CheckMate® 9900 CO_2/O_2 (Dansensor A/S, Ringsted, Denmark). VOC concentrations (ppb) were monitored with a SIFT-MS instrument (Voice 200, Syft Technologies Ltd, Christchurch, New Zealand) in accordance with the methodology of Kuuliala et al. (20). Briefly, 21 VOCs (Table 1) were selected on the basis of literature screening and quantified directly from the headspace of individual carrot packages. The headspace of each package was sampled for 300 s, resulting in ca. 40 datapoints. During sampling, the headspace was also connected to a bag (Nalophan™, Scentroid, Whitchurch-Stouffville, ON, Canada) filled with nitrogen to prevent package collapse. All measured concentrations (ppb) were corrected in accordance with the flow rate, measured with a Gilibrator 2 (Sensidyne, St. Petersburg, FL, USA) or a soap film flowmeter (Agilent Technologies, California, USA). Relative standard deviations ($SD_\%$) were calculated for the concentration of each VOC in each sample (individual carrot package) according to equation (1),

$$SD_\% = \frac{SD_m}{x_m} \times 100 \% \qquad (1)$$

where x_m is the average and SD_m the standard deviation during the scan. One primary product ion per VOC was selected based on optimizing the following criteria: 1) $SD_\% < 25\%$; 2) minimum number of mass conflicts; 3) high branching ratio; and 4) lowest concentration. In case of unavoidable mass conflicts (Table 1: C01, C20, C21), an average concentration was calculated.

Total psychrotrophic counts (TPC) were determined by classical plate counting techniques. From each individual carrot package, 20.0 ± 0.1 g was aseptically weighed into a stomacher bag, diluted with peptone physiological saline solution (PPS: 0.85 % m/v NaCl, 0.1 % m/v peptone), homogenized for 1 min and used for preparing decimal dilution series in PPS. Plate count agar (PCA; Oxoid Ltd., Basingstoke, Hampshire, UK) was inoculated with appropriate dilutions by pour plating and incubated at 22 °C for 3 days. After incubation, colony forming units (CFU) were enumerated and expressed on a logarithmic scale (log CFU g⁻¹).

Probabilistic Topic Modeling: General Principles

Latent dirichlet allocation – introduced by Blei et al. in 2003 (21) – represents one of the most popular contemporary topic modeling methods in the scientific literature. Concise descriptions of the general principles of LDA and its application in spoilage analysis are given below.

Because of LDA's background in natural language processing and text mining, text-associated terminology is often used for denoting its basic concepts. For this reason, LDA users typically refer to their data by the name *corpus*, meaning a collection of *documents* that consist of *words*. Here, a word refers to the basic unit of discrete data (21). It it should be noted, however, that the use of LDA is not limited to text analysis.

Table 1. VOCs and SIFT-MS parameters for the analysis of shredded carrot.

VOC[a]	ID[b]	P[c]	m/z[d]	b[e]	k[f]	Product ion
Alcohols						
Butanediol mix	C01					
1,2-Butanediol		NO+	89	100	2.3	$C_4H_9O_2^+$ [g]
		O_2^+	59	75	2.3	$C_3H_7O^+$
1,4-Butanediol		NO+	71	60	2.6	$C_4H_7O^+$
		NO+	89	40	2.6	$C_4H_9O_2^+$ [g]
		O_2^+	71	20	3.0	$C_4H_7O^+$
2,3-Butanediol		H_3O^+	91	100	3.0	$C_4H_{10}O_2.H^+$
		H_3O^+	109		2.6	$C_4H_{10}O_2.H^+.H_2O$
		NO+	89	100	2.3	$C_4H_9O_2^+$ [g]
Ethanol	C02	NO+	45	100	1.2	$C_2H_5O^+$ [g]
		NO+	63		1.2	$C_2H_5O^+.H_2O$
		NO+	81		1.2	$C_2H_5O^+.2(H_2O)$
Eugenol	C03	NO+	164	100	2.5	$C_{10}H_{12}O_2^+$ [g]
		O_2^+	164	100	2.5	$C_{10}H_{12}O_2^+$
Methanol	C04	H_3O^+	33	100	2.7	CH_5O^+ [g]
		H_3O^+	51		2.7	$CH_3OH.H^+.H_2O$
Aldehydes						
Acetaldehyde	C05	H_3O^+	45	100	3.7	$C_2H_5O^+$ [g]
		H_3O^+	63		3.7	$C_2H_5O^+.H_2O$
		H_3O^+	81		3.7	$C_2H_5O^+.2(H_2O)$
Octanal	C06	NO+	127	100	3.0	$C_8H_{15}O^+$ [g]
Amines						
Ammonia	C07	O_2^+	17	100	2.4	NH_3^+ [g]
Aniline	C08	H_3O^+	94	100	2.8	$C_6H_5NH_2.H^+$ [g]
Isopropyl amine	C09	H_3O^+	60	65	2.4	$(CH_3)_2CHNH_2.H^+$
		NO+	44	53	1.5	$C_2H_6N^+$
		NO+	58	47	1.5	$C_3H_8N^+$ [g]

Table 1. (Continued). VOCs and SIFT-MS parameters for the analysis of shredded carrot.

VOC[a]	ID[b]	P[c]	m/z[d]	b[e]	k[f]	Product ion
Esters						
Ethyl acetate	C10	H_3O^+	89	100	2.9	$CH_3COOC_2H_5.H^+$
		H_3O^+	107		2.9	$CH_3COOC_2H_5.H^+.H_2O$
		NO^+	118	90	2.1	$NO^+.CH_3COOC_2H_5$[g]
Ketones						
6-Methyl-5-hepten-2-one	C11	NO^+	108	40	2.5	$C_8H_{12}^+$
		NO^+	126	60	2.5	$C_8H_{14}O^+$[g]
		O_2^+	108	65	2.5	$C_8H_{12}^+$
		O_2^+	126	35	2.5	$C_8H_{14}O^+$
β-Ionone	C12	H_3O^+	193	100	3.0	$C_{13}H_{20}O.H^+$
		H_3O^+	211		3.0	$C_{13}H_{20}O.H^+.H_2O$
		NO^+	192	95	2.5	$C_{13}H_{20}O^+$[g]
Other						
2-Pentylfuran	C13	NO^+	138	100	2.0	$C_9H_{14}O^+$[g]
4-Isopropyl toluene	C14	NO^+	134	97	2.0	$C_{10}H_{14}^+$[g]
Coumarin	C15	NO^+	146	90	2.5	$C_9H_6O_2^+$[g]
		NO^+	147		2.5	$C_9H_6O_2.H^+$
Sulfur compounds						
2-Propanethiol	C16	H_3O^+	77	80	2.3	$C_3H_7SH_2^+$[g]
Dimethyl disulfide	C17	H_3O^+	95	100	2.6	$(CH_3)_2S_2.H^+$
		NO^+	94	100	2.4	$(CH_3)_2S_2^+$[g]
Ethyl mercaptan	C18	O_2^+	47	27	2.4	CH_3S^+
		O_2^+	62	53	2.4	$C_2H_5SH^+$[g]
Hydrogen sulfide	C19	H_3O^+	35	100	1.6	H_3S^+[g]
		O_2^+	34	100	1.4	H_2S^+
Terpenes						
Terpene mix 1	C20					
α-Humulene		NO^+	204	80	2.0	$C_{15}H_{24}^+$[g]
Trans-caryophyllene		NO^+	204	78	2.0	$C_{15}H_{24}^+$[g]

Table 1. (Continued). VOCs and SIFT-MS parameters for the analysis of shredded carrot.

VOC[a]	ID[b]	P[c]	m/z[d]	b[e]	k[f]	Product ion
Terpene mix 2	C21					
β-Pinene		NO+	136	89	2.1	$C_{10}H_{16}^{+}$[g]
γ-Terpinene		H_3O^+	137	81	2.6	$C_{10}H_{17}^{+}$
		NO+	136	75	2.1	$C_{10}H_{16}^{+}$[g]
Myrcene		H_3O^+	137	58	2.6	$C_{10}H_{17}^{+}$
		NO+	136	55	2.2	$C_{10}H_{16}^{+}$[g]
		O_2^+	92	69	2.2	$C_7H_8^{+}$
R-limonene		NO+	136	88	2.2	$C_{10}H_{16}^{+}$[g]
Terpinolene		H_3O^+	137	85	2.6	$C_{10}H_{17}^{+}$
		NO+	136	100	2.2	$C_{10}H_{16}^{+}$[g]

[a] Volatile organic compound (VOC). [b] Compound ID. [c] Precursor ion. [d] Mass-to-charge (m/z) ratio. [e] Branching ratio (%). [f] Reaction rate coefficient ($\times 10^{-9}$ molecule cm^{-3} s^{-1}). [g] Primary product ions selected for quantification.

When examining a corpus, differences in the frequencies of words can usually be observed both within and between documents. Certain words may be abundant (or rare, respectively) in all documents, while some other words may be characteristic to a limited set of documents. However, it can also be often seen that certain words are more likely to co-occur than others. In linguistic terms, it could be stated that these words belong to the same *topic*, meaning that they have been derived from the same underlying probability distribution over the vocabulary of the corpus. A document belonging to the corpus could thus be produced as follows (21):

1) Choose a distribution over topics,
2) Choose a topic from the chosen distribution,
3) Choose a word from the chosen topic,
4) If the document length has not yet been reached, return to step 2.

In summary, LDA can be used for extracting the aforementioned distributions and thus for uncovering underlying ("latent") information of the corpus.

Probabilistic Topic Modeling: Spoilage Analysis

Modeling using LDA was recently introduced as an exploratory and selective method for characterizing the volatilome of (sea)food stored under modified atmospheres (19). In this study, LDA was used for modeling the volatilome of salmon fillet portions that had been stored for up to 13 days under MAP conditions. Individual salmon samples from different days of storage were treated as "documents", individual VOCs as "words", and their concentrations as the frequencies of each "word" in each "document". Hence, the extracted "topics" represented underlying probability distributions over the quantified VOCs and were thus denoted as "VOC profiles" or, shortly, "profiles". The likelihood of drawing a given VOC from a given distribution indicates the relative abundance of the said VOC in the said profile. Finally, the profiles were interpreted by examining their correlation with quality decay. For example, profiles that were dominant during the early stages of storage time were interpreted to signify freshness, whereas profiles that showed positive correlation with storage time and sensory rejection were associated with spoilage.

The identification of the extracted profiles provides several advantages in terms of spoilage analysis. By examining the distribution of profiles within samples, the quality status of the samples could be characterized, for example, a given sample could be considered 80 % "fresh" and 20 % "spoiled". On the other hand, by examining the distribution of VOCs within the spoilage-associated profiles, potential spoilage indicators could be identified.

Probabilistic Topic Modeling: Methods

In the present study, topic modeling was performed with R 4.0.2 (22) in accordance with the previously developed methodology (19). The individual carrot packages were treated as "documents", VOCs as "words", and their rounded concentrations as "frequencies". Respectively, the extracted "topics" were denoted as "VOC profiles" or "profiles". First, leave-one-out cross-validation was performed for selecting the number of profiles (k=2,...,10), using the functions of the package **topicmodels** (23) and the following tunings: *method*=Gibbs, *iter*=1000, *burnin*=500, *thin*=1. Based on the obtained results, an LDA model with five profiles was chosen for spoilage characterization. Relevant distributions (profiles within samples and VOCs within profiles) were visualized using the function **tidy** from the package tidytext (24) and the package **ggplot2** (25) or Microsoft Excel. The extracted profiles were interpreted by examining the relationships between the profiles, storage time and microbial growth. Finally, potential spoilage indicators were identified by examining the distribution of VOCs within spoilage-associated profiles.

Results and Discussion

The evolution of the headspace gases (O_2/CO_2) and TPC is presented in Figure 1.

The initial atmosphere (19.5 ± 0.25 % O_2, 0.57 ± 0.25 % CO_2) was rapidly lost, as oxygen was depleted from the headspace already during the first few days. Simultaneously, a linear increase in CO_2 levels led to concentrations of 50 ± 6.3 % by the end of storage time. These two phenomena could be attributed to two main reasons, namely 1) the respiration of the packed product and 2) microbial growth and metabolism. Briefly, MPV consume O_2 throughout storage, meaning that enclosing them in a closed system will eventually lead to the generation of anaerobic conditions over time and thus fermentative spoilage processes that cause off-odors and off-flavors in the packed product (6). To prevent such undesired phenomena, the packaging system and storage conditions (in particular, the storage temperature) should be designed to minimize the deterioration caused by respiration. For example, ensuring sufficient O_2 availability in the package headspace is typically achieved by choosing packaging materials that allow continuous gas exchange with the surrounding atmosphere throughout storage time. However, improving MPV quality control calls not only for establishing optimal storage conditions, but also for improving the capability to detect potential problems in the supply chain. For this reason, the present study focused on VOCs that could be attributed to problems arising from insufficient mass transfer.

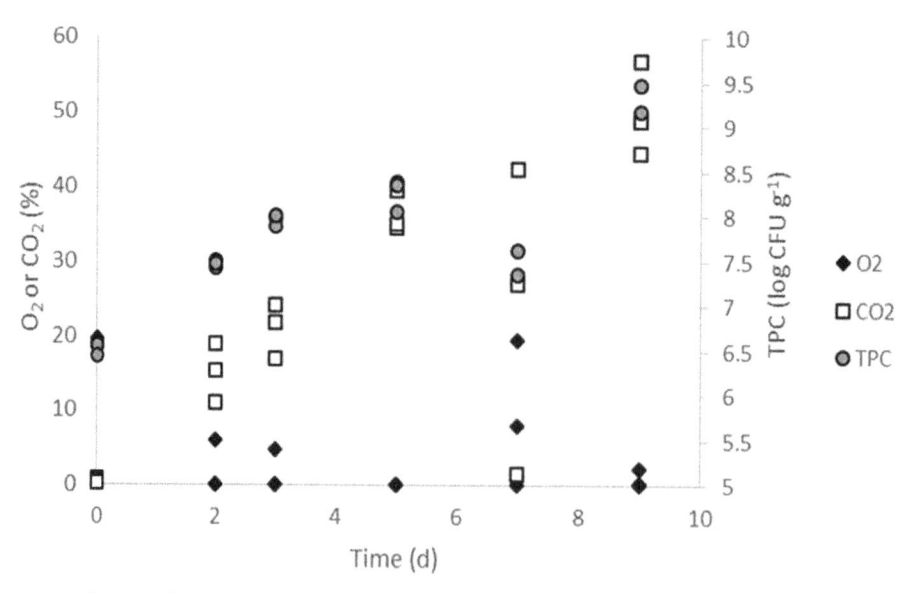

Figure 1. Evolution of headspace gases (O_2/CO_2) and total plate counts (TPC) as a function of storage time. The values of each individual sample (carrot package) are displayed as separate datapoints.

A rapid increase was also observed in TPC. Starting from 6.5 ± 0.07 log CFU g^{-1}, approximately 7.5 log CFU g^{-1} was reached by day 2 and over 9 log CFU g^{-1} by the end of the storage time. In the literature, the acceptability limit of the total microbial load of fresh-cut carrots ranges between ca. 7.0-8.0 log CFU g^{-1} (14, 26); in the present study, 8 log CFU g^{-1} was reached on day 3. It should be noted, however, that the point of consumer rejection could deviate from this moment. In the study of Vandekinderen et al. (14), the microbial shelf-life of untreated grated carrots stored at 7 °C was 3-4 days, whereas the odor and flavor were judged unacceptable at day 5. However, since the sensory quality of shredded carrot is affected by several parameters (including odor, flavor, color and appearance), the perceived odor may not always be the determining factor behind consumer rejection. Further experiments under different packaging and storage conditions could thus be helpful in elucidating the role of the carrot volatilome in consumer appreciation of MP carrots. The perplexities of the cross-validation models are listed in Table 2.

Perplexity indicates the capacity of the model to predict a new dataset that was not used for training the model (23). As shown in Table 2, an increasing value of k (number of extracted profiles) resulted in a slight decrease in perplexity and thus an increase in the performance of the model. When comparing the results obtained with each value of k (per column), it could be observed that while the lowest average perplexity was obtained with $k=7$ (2.8896 ± 1.3493), the standard deviations were relatively high. This could be attributed to the late-stage samples whose volatilome deviated from that of the early-stage samples (see also Figure 2): since the total number of early-stage samples was higher in the dataset, the prediction capacity was lower when a late-stage sample was used as the holdout set during cross-validation. On the other hand, when comparing the results obtained with each holdout sample (per row), the lowest value of k that never changed the (unrounded) perplexity by more than one standard deviation when compared to the minimizing k was $k=5$. Based on these results, an LDA model with five profiles was chosen for further analysis. It is important to emphasize that even though differences in prediction ability can be expected when using different holdout samples and/or values of k, no data should be discarded on this basis. In the present study, the full dataset was used for all modeling activities. The distribution of the profiles in the samples is shown as a function of storage time in Figure 2.

Table 2. The perplexities of the cross-validation models.

ID[a]	k[b]=2	k=3	k=4	k=5	k=6	k=7	k=8	k=9	k=10	SD[c]
0A	2.41	2.40	2.34	2.28	2.28	2.28	2.28	2.28	2.28	0.06
0B	2.48	2.47	2.40	2.34	2.34	2.34	2.34	2.34	2.34	0.06
0C	2.47	2.42	2.41	2.35	2.35	2.35	2.35	2.35	2.35	0.04
2A	2.24	2.27	2.23	2.17	2.17	2.17	2.17	2.17	2.17	0.04
2B	1.76	1.75	1.75	1.74	1.74	1.74	1.74	1.74	1.74	0.01
2C	2.52	2.50	2.43	2.34	2.34	2.34	2.34	2.34	2.34	0.08
3A	1.76	1.76	1.76	1.75	1.74	1.75	1.74	1.74	1.74	0.01
3B	1.57	1.53	1.49	1.48	1.48	1.48	1.48	1.48	1.48	0.03
3C	1.55	1.52	1.46	1.43	1.43	1.43	1.43	1.43	1.43	0.05
5A	2.95	2.57	2.48	2.47	2.47	2.47	2.47	2.47	2.47	0.16
5B	2.78	2.42	2.35	2.33	2.33	2.33	2.32	2.32	2.32	0.15
5C	5.46	5.20	5.20	5.02	4.96	4.96	4.96	4.96	4.96	0.18
7A	4.49	3.93	3.58	3.58	3.58	3.57	3.57	3.57	3.57	0.31
7B	3.01	2.81	2.80	2.79	2.79	2.79	2.79	2.79	2.78	0.07
7C	3.79	3.39	3.38	3.38	3.38	3.38	3.37	3.37	3.37	0.14
9A	3.20	3.07	3.07	3.06	3.06	3.06	3.06	3.06	3.06	0.05
9B	6.98	5.94	6.04	5.67	5.65	5.65	5.61	5.61	5.57	0.45
9C	8.04	8.23	7.24	6.14	5.99	5.94	6.54	6.56	6.54	0.85
Av[d]	3.3	3.12	3.02	2.91	2.89	2.89	2.92	2.92	2.92	
SD[c]	1.84	1.74	1.59	1.39	1.36	1.35	1.43	1.43	1.43	

[a] Holdout sample ID. [b] Number of extracted profiles. [c] Standard deviation. [d] Average.

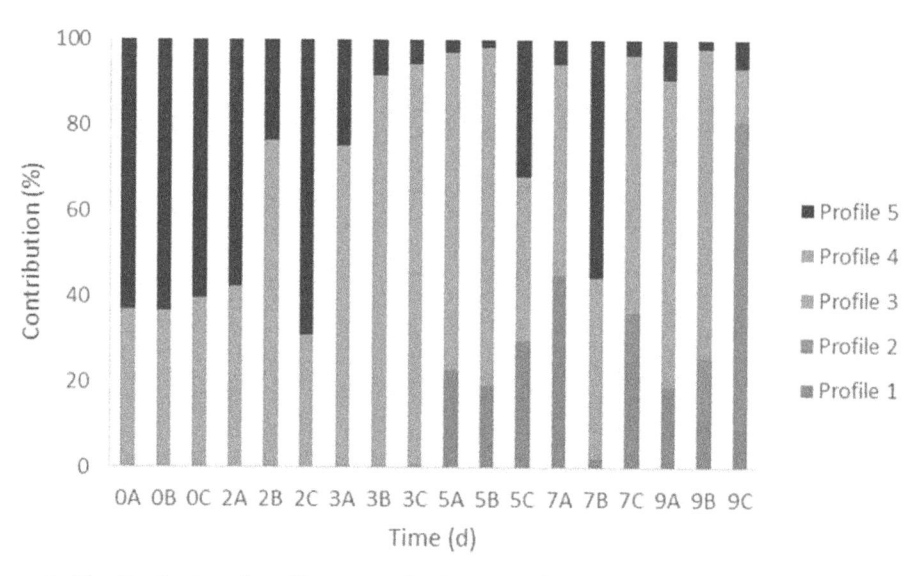

Figure 2. The distribution of profiles in samples (carrot packages) as a function of storage time. Each individual sample is displayed as a separate column, where the letter A-C indicates the daily replicate.

During the earliest days of storage (days 0-3), profiles 4 and 5 were dominant in all studied samples. The contribution of profile 5 was greatly reduced from day 3 on, whereas profile 4 was prominent until the last day of storage. Profile 3 had a low initial contribution (<3 %) and was prominent in some late-stage samples, but did not exhibit a clear trend. On the other hand, an increasing trend in the contributions of profiles 1 and 2 was observed from day 3-5 onwards, coinciding not only with the generation of anaerobic conditions but also with TPC levels exceeding 8 log CFU g^{-1} (see Figure 1). This is in line with several previous studies where ca. 7-7.5 log CFU g^{-1} has been identified as the onset point of prominent VOC production due to microbial metabolism (20, 27). Hence, profiles 4 and 5 could be associated with freshness and profiles 1 and 2 with spoilage in a closed packaging system.

Still, it should be noted that some fluctuation in profile patterns was observed as a function of storage time. This could be attributed not only to natural variation between individual carrot samples, but also to factors arising from the packaging and storage conditions. For example, sample 7B had considerably lower overall concentration levels when compared to other late-stage samples: this was most likely due to insufficient top-film sealing that prevented the formation of an anaerobic atmosphere (see also Figure 1). On the other hand, the impact of the headspace volume should be considered. Even though non-respiring products like meat and fish are often packed in rigid consumer trays in combination with a low-barrier top film, a respective approach for MPV could lead to package failure due to CO_2 buildup during storage time. While packing in bags reduces this risk, it simultaneously allows for a larger change in headspace volume and may thus lead to a dilution effect. However, although sample dilution may interfere with VOC trend detection, it simultaneously represents the actual scenario that a given quality monitoring solution (e.g., a VOC sensor) would face. The distribution of the top eight VOCs in the profiles is shown in Figure 3.

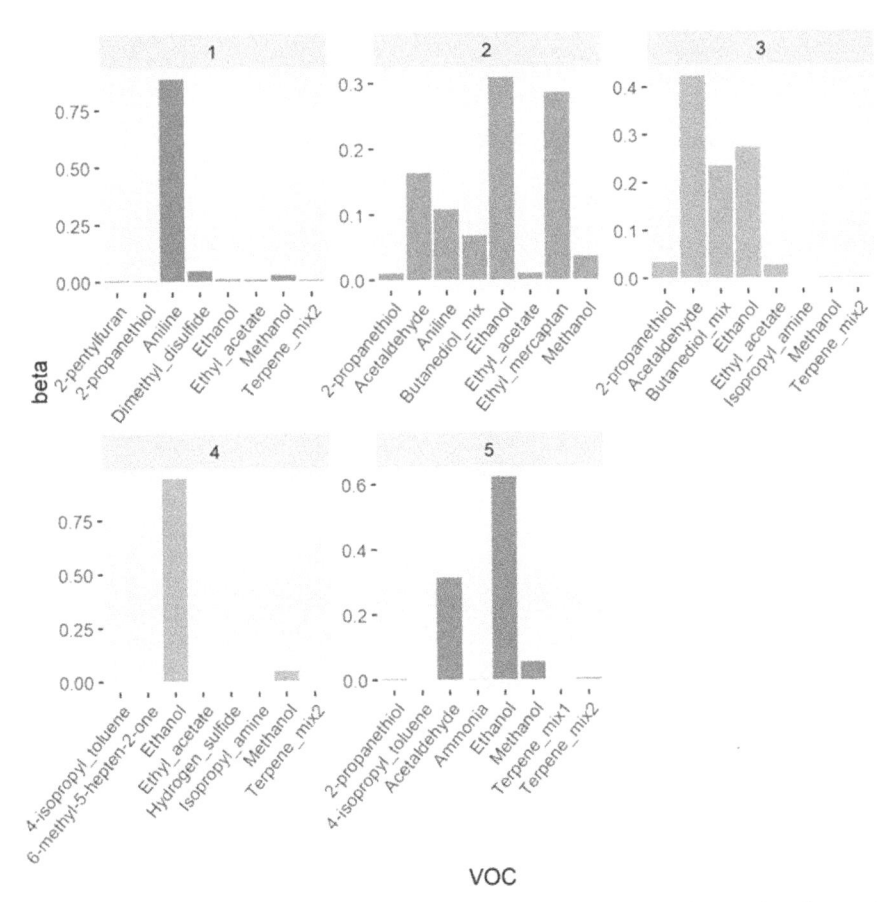

Figure 3. The distribution of top eight VOCs in the extracted profiles. Beta indicates the relative abundance (0<beta<1) of a given VOC in a given profile.

The freshness-associated profiles 4 and 5 were largely dominated by ethanol. Apart from methanol (both profiles) and acetaldehyde (profile 5), none of the other top eight VOCs exceeded 0.01 relative abundance in these profiles. However, this was due to the high levels of the aforementioned three compounds when compared to the other quantified VOCs. As previously highlighted (6), the olfactory threshold (OT) and offensiveness of different VOCs may greatly differ. Consequently, a high number of VOCs can be expected to contribute to the fresh aroma even when present in low quantities. For example, many of those compounds that were potentially involved in the second terpene mix (R-limonene, gamma-terpinene, myrcene, terpinolene) have previously been associated with the characteristic aroma of carrot or carrot-based products (17, 28, 29). Because of multiple mass conflicts (see Table 1), complementary experiments would be needed to enhance the distinction of these closely related compounds.

A more diverse VOC distribution could be observed in the spoilage-associated profiles (i.e., profiles 1 and 2). While all of the top eight VOCs of these three profiles exceeded 0.005 relative abundance, many of them were simultaneously prominent in the freshness-associated profiles (i.e., profiles 4 and 5). For example, ethanol and methanol were initially present in high levels (52600 ± 5800 and 4000 ± 700 ppb on day 0, respectively) but showed relatively little change during storage time. In contrast, those top eight VOCs that did not overlap with the freshness-associated profiles (aniline, dimethyl disulfide, ethyl mercaptan, 2-propanethiol and the alcohol mix) showed increasing trends as a function of storage time. The evolution of these compounds is shown as a function of time and TPC in Figure 4.

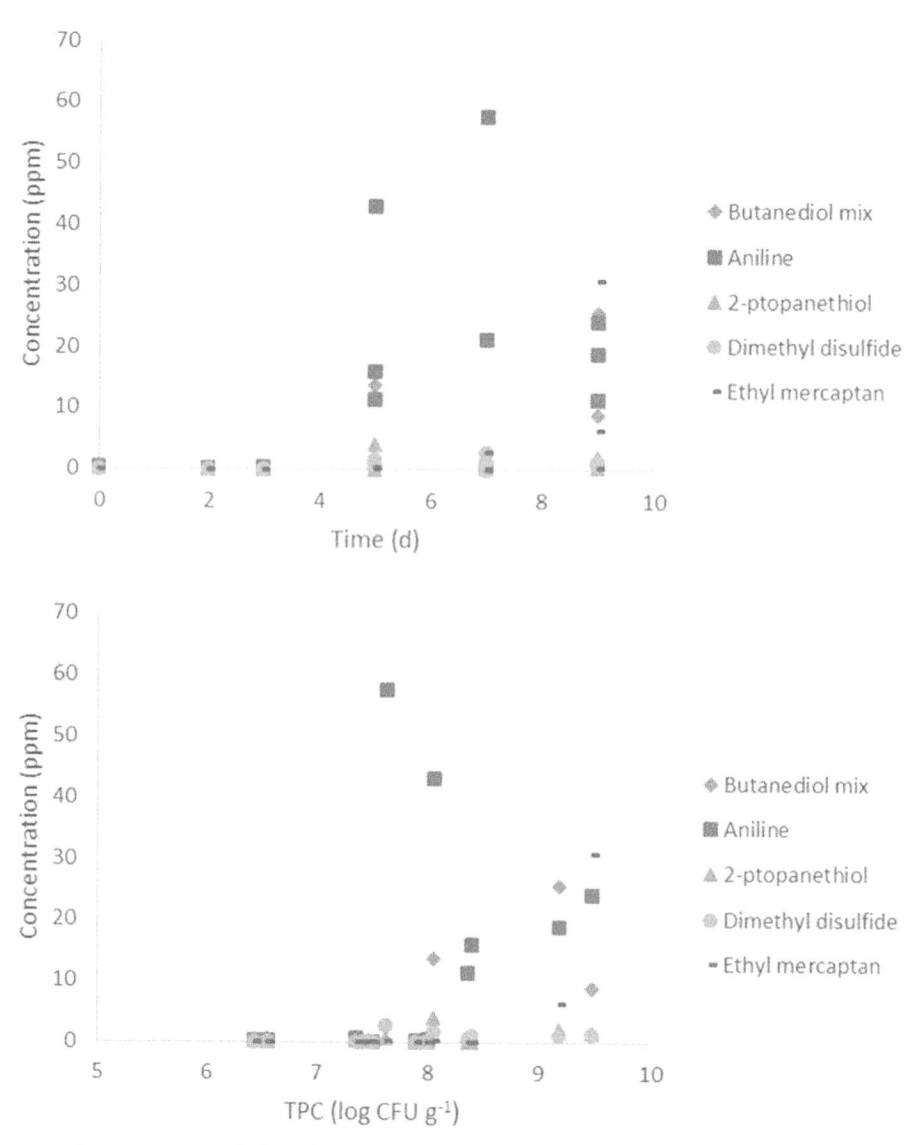

Figure 4. Concentrations of selected VOCs (ppm) as a function of storage time and total plate counts (TPC).

Ethanol is a well-known fermentation product in several food products stored under modified atmosphere conditions, including carrots (*30*). In previous studies focusing on MP carrots, growth of lactic acid bacteria (LAB) has commonly been observed (*31, 32*). Psychrotrophic LAB are able to grow in the absence of oxygen and have been found capable of producing several VOCs, including ethanol and 2,3-butanediol (*33*). While the production of sulfuric compounds has been most frequently associated with the microbiological spoilage of meat and fish, increasing levels have also frequently been observed over storage time in fresh produce (*34–36*). In contrast, the detection of high aniline levels is likely due to a mass conflict with another yet unidentified compound. However, it should be noted that in the case of MPVs, off-odors may be due not only to microbial metabolism, but also to plant metabolism (6). Further experiments would thus be needed to investigate the production mechanisms of these compounds in MP carrots, as well as to evaluate their role under different packaging and storage conditions.

The identification of spoilage-related VOCs calls for establishing well-justified identification criteria. Firstly, VOCs that are initially present at low levels in the food product and exhibit an increasing trend over storage time have been typically considered most useful spoilage indicators in practical settings. The results of the present study indicate that despite the presence of several abundant VOCs, relatively few compounds could be identified under the studied conditions. However, it should be noted that further experiments would be beneficial in specifying the identity of closely related compounds. Secondly, exceeding the OT has frequently been considered important, as VOCs that remain below their OT cannot be perceived by the human nose. As previously discussed in several studies (6, 19), however, the application of OTs in spoilage analysis poses several challenges. On the one hand, previously reported OTs have rarely been determined in complex VOC mixtures, meaning that the reported values do not directly indicate whether a VOC can be perceived as a part of the food volatilome. On the other hand, the capacity to detect and quantify VOCs below their OT could greatly enhance the detection of early spoilage. For these reasons, the concept "potential spoilage indicator" has been established to refer to VOCs that could be useful for quality monitoring purposes, irrespective of their OT (19).

Thanks to the increasing volume and complexity of modern VOC datasets, multivariate statistical methods are greatly in demand. LDA is a flexible method that imposes few assumptions and shows promising potential in a variety of biological applications. Because of its unsupervised nature, additional data is necessary for interpreting the extracted profiles, yet the lack of a dependent variable reduces the risk of incorrectly assuming a (linear) relationship between the studied variables, which could potentially lead to misleading conclusions about deterioration mechanisms when using methods like partial least squares regression (PLS) analysis (19).

Conclusions

In order to assess the spoilage of food products under commercially relevant and realistic conditions, non-destructive measurement techniques can be considered highly beneficial. In the present study, the combination of real-time mass spectrometry and probabilistic topic modeling allowed for the identification of two major types of VOCs that contributed to the spoilage volatilome under the studied conditions: 1) abundant but stable (e.g., ethanol) and 2) increasing over storage time (e.g., dimethyl disulfide). While these results still represent a preliminary view into carrot spoilage, the developed methodology could be used for studying the volatilome under different packaging and storage conditions. In particular, further experiments under equilibrium modified atmosphere packaging (EMAP) systems could be expected to be of interest to both scientific and industrial communities.

Acknowledgments

L. Kuuliala acknowledges the support of the Research Foundation Flanders for a Junior Post-doctoral Fellow (1222020N). The data collection was realized in the framework of the TERAFOOD project supported by the European Regional Development Fund and the province Oost-Vlaanderen, the "Onderzoeksprogramma Artificiële Intelligentie (AI) Vlaanderen" program supported by the Flemish Government.

References

1. Francis, G. A.; Thomas, C.; O'beirne, D. The microbiological safety of minimally processed vegetables. *Int. J. Food Sci. Tech.* **1999**, *34* (1), 1–22.
2. Ragaert, P.; Devlieghere, F.; Debevere, J. Role of microbiological and physiological spoilage mechanisms during storage of minimally processed vegetables. *Postharvest Biol. Technol.* **2007**, *44* (3), 185–194.
3. Rosberg, A. K.; Darlison, J.; Mogren, L.; Alsanius, B. W. Commercial wash of leafy vegetables do not significantly decrease bacterial load but leads to shifts in bacterial species composition. *Food Microbiol.* **2021**, *94*, 103667.
4. Gao, J.; Jang, H.; Huang, L.; Matthews, K. R. Influence of product volume on water antimicrobial efficacy and cross-contamination during retail batch washing of lettuce. *Int. J. Food Microbiol.* **2020**, *323*, 108593.
5. Huang, K.; Tian, Y.; Tan, J.; Salvi, D.; Karwe, M.; Nitin, N. Role of contaminated organic particles in cross-contamination of fresh produce during washing and sanitation. *Postharvest Biol. Technol.* **2020**, *168*, 111283.
6. Ioannidis, A.; Kerckhof, F.; Riahi Drif, Y.; Vanderroost, M.; Boon, N.; Ragaert, P.; De Meulenaer, B.; Devlieghere, F. Characterization of spoilage markers in modified atmosphere packaged iceberg lettuce. *International Journal of Food Microbiology* **2018**, *279*, 1–13.
7. Hu, X.; Huang, E.; Barringer, S. A.; Yousef, A. E. Factors affecting Alicyclobacillus acidoterrestris growth and guaiacol production and controlling apple juice spoilage by lauric arginate and ε-polylysine. *LWT* **2020**, *119*, 108883.
8. Zhang, B.; Samapundo, S.; Pothakos, V.; de Baenst, I.; Sürengil, G.; Noseda, B.; Devlieghere, F. Effect of atmospheres combining high oxygen and carbon dioxide levels on microbial spoilage and sensory quality of fresh-cut pineapple. *Postharvest Biol. Technol.* **2013**, *86*, 73–84.
9. Frank, D.; Piyasiri, U.; Archer, N.; Jenifer, J.; Appelqvist, I. Influence of saliva on individual in-mouth aroma release from raw cabbage (Brassica oleracea var. capitata f. rubra L.) and links to perception. *Heliyon* **2018**, *4* (12), e01045.
10. Raseetha, S.; Heenan, S. P.; Oey, I.; Burritt, D. J.; Hamid, N. A new strategy to assess the quality of broccoli (Brassica oleracea L. italica) based on enzymatic changes and volatile mass ion profile using Proton Transfer Reaction Mass Spectrometry (PTR-MS). *Innov. Food Sci. Emerg. Technol.* **2011**, *12* (2), 197–205.
11. Lavelli, V.; Pagliarini, E.; Ambrosoli, R.; Minati, J. L.; Zanoni, B. Physicochemical, microbial, and sensory parameters as indices to evaluate the quality of minimally-processed carrots. *Postharvest Biol. Technol.* **2006**, *40* (1), 34–40.
12. Fai, A. E. C.; Alves de Souza, M. R.; de Barros, S. T.; Bruno, N. V.; Ferreira, M. S. L. Gonçalves,Édira Castello Branco de Andrade Development and evaluation of biodegradable films and coatings obtained from fruit and vegetable residues applied to fresh-cut carrot (Daucus carota L.). *Postharvest Biol. Technol.* **2016**, *112*, 194–204.
13. Alegria, C.; Pinheiro, J.; Gonçalves, E. M.; Fernandes, I.; Moldão, M.; Abreu, M. Evaluation of a pre-cut heat treatment as an alternative to chlorine in minimally processed shredded carrot. *Innov. Food Sci. Emerg. Technol.* **2010**, *11* (1), 155–161.
14. Vandekinderen, I.; Devlieghere, F.; Van Camp, J.; Denon, Q.; Alarcon, S. S.; Ragaert, P.; De Meulenaer, B. Impact of a decontamination step with peroxyacetic acid on the shelf-life,

15. Cozzolino, R.; De Giulio, B.; Pellicano, M. P.; Pace, B.; Capotorto, I.; Martignetti, A.; D'Agresti, M.; Laurino, C.; Cefola, M. Volatile, quality and olfactory profiles of fresh-cut polignano carrots stored in air or in passive modified atmospheres. *LWT* **2021**, *137*, 110408.

16. Condurso, C.; Cincotta, F.; Tripodi, G.; Merlino, M.; Giarratana, F.; Verzera, A. A new approach for the shelf-life definition of minimally processed carrots. *Postharvest Biol. Technol.* **2020**, *163*, 111138.

17. Negri Rodríguez, L. M.; Arias, R.; Soteras, T.; Sancho, A.; Pesquero, N.; Rossetti, L.; Tacca, H.; Aimaretti, N.; Rojas Cervantes, M. L.; Szerman, N. Comparison of the quality attributes of carrot juice pasteurized by ohmic heating and conventional heat treatment. *LWT* **2021**, *145*, 111255.

18. Koutidou, M.; Grauwet, T.; Acharya, P. Effect of different combined mechanical and thermal treatments on the volatile fingerprint of a mixed tomato–carrot system. *J. Food Eng.* **2016**, *168*, 137–147.

19. Kuuliala, L.; Pérez-Fernández, R.; Tang, M.; Vanderroost, M.; De Baets, B.; Devlieghere, F. Probabilistic topic modelling in food spoilage analysis: A case study with Atlantic salmon (Salmo salar). *Int. J. Food Microbiol.* **2021**, *337*, 108955.

20. Kuuliala, L.; Sader, M.; Solimeo, A.; Pérez-Fernández, R.; Vanderroost, M.; De Baets, B.; De Meulenaer, B.; Ragaert, P.; Devlieghere, F. Spoilage evaluation of raw Atlantic salmon (Salmo salar) stored under modified atmospheres by multivariate statistics and augmented ordinal regression. *Int. J. Food Microbiol.* **2019**, *303*, 46–57.

21. Blei, D. M.; Ng, A. Y.; Jordan, M. I. Latent Dirichlet Allocation. *J. Mach. Learn. Res.* **2003**, *3*, 993–1022.

22. R Core Team. *R: A language and environment for statistical computing*; R Foundation for Statistical Computing: Vienna, Austria, 2020; URL: https://www.R-project.org/.

23. Grün, B.; Hornik, K. topicmodels: An R Package for Fitting Topic Models. *J. Stat. Softw.* **2011**, *40* (13), 1–30.

24. Silge, J.; Robinson, D. tidytext: Text Mining and Analysis Using Tidy Data Principles in R. *J. Open Source Softw.* **2016**, *1* (3)

25. Wickham, H. *ggplot2: Elegant Graphics for Data Analysis*; Springer-Verlag: New York, 2016.

26. Piscopo, A.; Zappia, A.; Princi, M. P.; De Bruno, A.; Araniti, F.; Antonio, L.; Abenavoli, M. R.; Poiana, M. Quality of shredded carrots minimally processed by different dipping solutions. *J. Food Sci. Technol.* **2019**, *56*, 2584–2593.

27. Kuuliala, L.; Al Hage, Y.; Ioannidis, A.-G..; Sader, M.; Kerckhof, F.-M.; Vanderroost, M.; Boon, N.; De Baets, B.; De Meulenaer, B.; Ragaert, P.; Devlieghere, F. Microbiological, chemical and sensory spoilage analysis of raw Atlantic cod (Gadus morhua) stored under modified atmospheres. *Food Microbiol.* **2018**, *70*, 232–244.

28. Keskin, M.; Guclu, G.; Sekerli, Y. E.; Soysal, Y.; Selli, S.; Kelebek, H. Comparative assessment of volatile and phenolic profiles of fresh black carrot (Daucus carota L.) and powders prepared by three drying methods. *Sci. Hortic.* **2021**, *287*, 110256.

29. Keser, D.; Guclu, G.; Kelebek, H.; Keskin, M.; Soysal, Y.; Sekerli, Y. E.; Arslan, A.; Selli, S. Characterization of aroma and phenolic composition of carrot (Daucus carota 'Nantes')

30. Hisashi, K.-N.; Watada, A. E. Effects of Low-oxygen Atmosphere on Ethanolic Fermentation in Fresh-cut Carrots. *J. Am. Soc. Hortic. Sci.* **1997**, *122*, 107–111.
31. Emmambux, N. M.; Minnaar, A. The effect of edible coatings and polymeric packaging films on the quality of minimally processed carrots. *J. Sci. Food Agric.* **2003**, *83*, 1065–1071.
32. Nguyen-the, C.; Carlin, F. The microbiology of minimally processed fresh fruits and vegetables. *Crit. Rev. Food Sci. Nutr.* **1994**, *34* (4), 371–401.
33. Pothakos, V.; Nyambi, C.; Zhang, B.; Papastergiadis, A.; De Meulenaer, B.; Devlieghere, F. Spoilage potential of psychrotrophic lactic acid bacteria (LAB) species: Leuconostoc gelidum subsp. gasicomitatum and Lactococcus piscium, on sweet bell pepper (SBP) simulation medium under different gas compositions. *Int. J. Food Microbiol.* **2014**, *178*, 120–129.
34. Chen, H.; Zhang, M.; Guo, Z. Discrimination of fresh-cut broccoli freshness by volatiles using electronic nose and gas chromatography-mass spectrometry. *Postharvest Biol. Technol.* **2019**, *148*, 168–175.
35. Du, X.; Chen, H.; Zhang, Z.; Qu, Y.; He, L. Headspace analysis of shelf life of postharvest arugula leaves using a SERS-active fiber. *Postharvest Biol. Technol.* **2021**, *175*, 111410.
36. Spadafora, N. D.; Amaro, A. L.; Pereira, M. J.; Müller, C. T.; Pintado, M.; Rogers, H. J. Multi-trait analysis of post-harvest storage in rocket salad (Diplotaxis tenuifolia) links sensorial, volatile and nutritional data. *Food Chem.* **2016**, *211*, 114–123.

Beginning of the reference list continues from prior page:

powders obtained from intermittent microwave drying using GC–MS and LC–MS/MS. *Food Bioprod. Process.* **2020**, *119*, 350–359.

Chapter 10

Real-Time Monitoring of Flavoring Starter Cultures for Different Food Matrices Using PTR-MS

Vittorio Capozzi,[1] Mariagiovanna Fragasso,[2] Iuliia Khomenko,[3] Patrick Silcock,[4] and Franco Biasioli[3,*]

[1]National Research Council—Institute of Sciences of Food Production (ISPA) c/o CS-DAT, via Michele Protano, 71121 Foggia, Italy

[2]Department of Agriculture, Food, Natural Resources, and Engineering, University of Foggia, via Napoli 25, 71121 Foggia, Italy

[3]Research and Innovation Centre, Fondazione Edmund Mach, via E. Mach 1, 38098 San Michele all'Adige (TN), Italy

[4]Department of Food Science, University of Otago, PO Box 56, Dunedin 9054, New Zealand

*Email: franco.biasioli@fmach.it

In fermented foods, volatile organic compounds (VOCs), which are often metabolic products of microorganisms, form a subset of the chemical compounds contributing to the sensory perceptions arising during product consumption. Direct injection mass spectrometry (DIMS) techniques allow for the direct and real-time measurement of VOC release without the need for laborious sample treatment, extraction procedures, and chromatographic separation. DIMS has been successfully applied in different fields, including characterizing the formation of flavor compounds associated with fermentation bioprocesses in foods and beverages. In this chapter, following a general overview on DIMS for the study of fermentation generated VOCs, the use of proton transfer reaction time-of-flight mass spectrometry (PTR-TOF-MS to investigate VOC generation during the fermentation bioprocesses in yeast-based fermentations and to characterize the flavor contributions of starter cultures is described. To this extent, a panel of five experiments is presented that demonstrates a pipeline of increasing complexity where DIMS is used to monitor VOC release in association with a) yeasts grown in a standard culture media, b) the growth of different yeast starter cultures in a real food matrix, c) the interaction between yeast starter cultures and different wheat flours, d) the interaction between different combinations of starter cultures grown in multiple food matrices, and e) the contribution of commercial starter cultures to specific flavor attributes in a food matrix. Fermentation is globally recognized as one of the key sustainable technologies in food science, and DIMS offers a low-cost, time-saving, and low-impact methodology to investigate fundamental themes and support agroindustry applications. Hence, the proposed approach is of

© 2021 American Chemical Society

interest to the fields of food biotechnology and flavor science but, this coupling of 'green' biotechnological and analytical solutions also represents a bridge towards improved sustainability of agro-food systems.

Introduction

Volatile organic compounds (VOCs) are organic compounds that are released as gases at room temperature owing to their physical properties. Across many disciplines, from atmospheric chemistry to plant physiology to food chemistry, VOCs play a key role and, in particular, in food where VOCs make an important contribution to the perceived quality of food (1–3). In analytical chemistry, forming a subset within the field of mass spectrometry (MS) is a family of VOC monitoring strategies that fall under the umbrella of direct injection mass spectrometry (DIMS) (1). These DIMS techniques comprise *inter alia*, proton transfer reaction-mass spectrometry (PTR-MS), atmospheric pressure chemical ionization-mass spectrometry (APCI-MS), selected ion flow tube-mass spectrometry (SIFT-MS), ion mobility mass spectrometry (IMMS) and MS-based electronic noses (MS-e-nose) (4). The common denominator of these techniques is the direct injection of headspace gases into the ionization chamber with no sample treatment, VOC extraction, or chromatographic separation (1, 4–6). The main differences in the analytical approaches are summarized in Table 1.

Table 1. Principal features characterizing some of the main DIMS analytical approaches (1).

DIMS technique	
Discontinuous sample injection	MS-e-nose
Continuous sample injection	
Precursor ions from sampled air	APCI-MS
Precursor ions from a controlled source	
Largely without precursor ions selection	PTR-MS
With precursor ions selection	SIFT-MS

DIMS techniques are versatile tools for understanding VOC release in agri-food applications, combining low-cost analytical strategies, fast sample processing and good analytical performances (1, 4–6). In fact, these techniques find multipurpose applications in the food chain, from high-throughput sample screening to process monitoring (4, 5, 7, 8). In particular, the non-destructive character of these analytical approaches allows the online monitoring of VOCs associated with given systems (1, 4). Experimentally it is feasible to follow time-dependent phenomena in agro-food systems, such as fermentations, shelf-life changes and thermal treatments.

About one-third of foodstuffs consumed worldwide are fermented foods and beverages (9), an assorted class of edible matrices "produced through controlled microbial growth, and the conversion of food components through enzymatic action" (10). This category includes the fermentation of cereals, vegetables, bamboo shoots, legumes, roots/tubers, milk, meat, fish and also alcoholic beverages (11). The fermentation of this wide range of matrices involves several genera/species of bacteria, yeasts, and filamentous fungi (11). The diversity of microorganisms that may be used to produce the fermented foods and beverages affects the main facets of quality, modulating nutritional, functional and sensory properties, and the general safety of the products (*12*). In order to assure high quality and safety in the modern agro-food systems, starter cultures have become widely used. Starter cultures are "A microbial preparation of large numbers of cells of at least one microorganism" of adequate quality and quantity "to be added to a raw material to produce a fermented food by accelerating and steering its fermentation process" (*13*). Microorganisms, which are metabolically active during fermentation, produce many VOCs as secondary metabolites, generally termed microbial VOCs (mVOCs) (*3*). Notably, the study of VOCs from fermented foods and beverages includes the contributions of the microorganisms' volatilome to that particular fermentation process (*14–17*). This aspect is relevant in the design and selection of the appropriate cultures that are to be inoculated to enable desired flavor characteristics to be achieved (*18–20*).

In this chapter, after a general overview of the DIMS application in fermented matrices, PTR-MS is used as a model instrumental technique to demonstrate the potential for online monitoring of fermentative bioprocesses and to enable the tailoring of flavor starters. In particular, based on published studies, the analytical potential of PTR-time-of-flight-MS (PTR-TOF-MS) is illustrated by way of its characterization of flavor-generating yeast cultures, monitoring headspace VOCs during a) yeast strains grown in a standard culture media, b) the growth of different yeast starter cultures in a real food matrix, c) the interaction between yeast starter cultures and different wheat flours on VOC release, d) the interaction between different combinations of starter cultures grown in multiple food matrices, and e) to investigate the basis for culture manufacturer's claims on the contribution of commercial starter cultures to specific flavor attributes in a food matrix.

Experimental Methods

Preparation of Samples with Microbiological Resources/Starter Cultures

Study a): After an overnight pre-culture at 28 °C in liquid Yeast Extract–Peptone–Dextrose (YPD) the six yeast strains were inoculated onto 2 mL solid YPD medium (YPD supplemented with 2% agar) inside a 20 mL vial and closed with a screw cap with a silicon/PTFE septum (*21*).

Study b): The American Association of Cereal Chemists AACC 10-10B procedure was applied with minor modifications to prepare the dough and bread samples. The basic dough preparations were inoculated with different commercial microbial preparations. The amount of each yeast preparation was added in accordance with the manufacturer's recommendations for each yeast (*22*).

Study c): Following the dough and baking procedure described in study b, a full factorial design of two starter cultures (Y1, Y2) and four flour types (F1-F4) was prepared for testing (*23*).

Study d): Two *Saccharomyces cerevisiae* strains and two non-*Saccharomyces* strains (*Metschnikowia pulcherrima* and *Torulaspora delbrueckii*) were inoculated into a commercial red grape juice and a red grape must (Apulian autochthonous grape variety 'Uva di Troia', 20° Babo; 7.2 g/L total acidity; 3.5 g/L malic acid; pH 3.5; free ammonium 163.5 mg/L) to reach the final concentration of 1×10^6 and 1×10^4 colony-forming units per milliliter (CFU/mL) for non-*Saccharomyces* and *S. cerevisiae*, respectively (*24*).

Study e): According to the manufacturers' specifications, dough samples were inoculated each with one of four different commercial yeast preparations, one control culture and three cultures with specific flavor traits (*19*).

With the exception of two *S. cerevisiae* laboratory strains (study a) (available in culture collections) and two wine strains (studies a and d) (available from the authors), all the microorganisms tested in the study were sub-cultured from commercial starter cultures.

Only essential details have been reported for the case studies a-e. For the complete procedures followed for the preparation of the experimental samples, please refer to the cited literature.

Common Details about the Instrumental Analysis Performed with PTR-TOF-MS

In general, for all reported experiments, VOCs released by the sample were measured by directly connecting the sample's headspace for a limited time, typically 60 s, to the drift tube of the PTR-MS instrument where ionization takes place. All experiments described were performed with a PTR-TOF 8000 apparatus (Ionicon Analytik GmbH, Innsbruck, Austria). The instrumental parameters in the drift tube were as follows: drift voltage 557 V, drift temperature 110 °C, drift pressure 2.30 mbar, generating an E/N value of about 140 Townsend (1 Td = 10^{-17} V cm^2) (E corresponds to the electric field and N to the gas number density). The sampling time per acquisition channel was 0.1 ns, amounting to 350,000 channels per mass spectrum up to m/z 400. Every single spectrum is the sum of 28,600 acquisitions, each lasting 35 µs, resulting in an analytical time resolution of 1 s. At least a 1 min interval was maintained between measurements to avoid memory effects. Measurements were automated by using a multipurpose gas chromatography (GC) automatic sampler (Autosampler, Gerstel GmbH, Mulheim an der Ruhr, Germany) as described elsewhere (4). A gas calibration unit (GCU, Ionicon Analytik GmbH, Innsbruck, Austria) was employed to generate zero air for flushing sample headspace in order to prevent the microbiological and chemical contamination of a sample headspace. In fermentation systems where high ethanol concentrations were present in the sample headspace, an argon dilution system was applied after headspace sampling to reduce primary ion depletion and ethanol cluster formation (25). The dilution ratio was one part headspace to three parts argon. The argon flow rate was 120 sccm and was controlled by a multigas controller (MKS Instruments, Inc). The following data processing was carried out: *i)* the count losses due to the ion detector dead-time were rectified by adopting applications of Poisson statistics; *ii)* compound annotation was performed to match the data with fragmentation of reference standards and with information reported in the scientific literature; *iii)* noise reduction, baseline removal and peak intensity extraction were carried out using modified Gaussians to fit the peaks; *iv)* data analysis was performed by principal component analysis (PCA), analysis of variance (ANOVA), Tukey's post-hoc test, and other statistical tests, adjusting existing packages created in R. For more details about data mining, including internal calibration to achieve a mass accuracy (up to 0.001 Th) and calculation of peak intensity in ppbv, please refer to the information reported in the literature (4).

Results and Discussion

Different instrumental approaches that belong to the category of DIMS techniques have found application in experimental plans that look to understand VOC release during fermentation across a range of matrices. Table 2 summarizes some examples of studies published over the last ten years, in which the potential of PTR-TOF-MS methods to study VOCs associated with fermented foods and beverages has been demonstrated.

PTR-TOF-MS, in particular, has found extensive use in these studies, including assessing of the effect of main categories of commercial starter cultures, i.e., lactic acid bacteria and saccharomycetes. Using saccharomycetes as model organism, representing an important subset of starter cultures, and PTR-TOF-MS as a model analytical technique, five experiments are presented that demonstrate the potential of real-time flavor compound measurements by DIMS to investigate to role of starter cultures in flavor development across different food (or food-like) matrices.

Table 2. PTR-TOF-MS applied to the detection of VOCs in fermented foods and beverages over the past decade.

Fermented product	Application
Cocoa beans	Study of fermented cocoa beans from diverse geographical origins (26)
Bread	Investigation of the VOCs released using different yeast starter cultures in breadmaking (22)
	Study of the impact of flour, yeast and their interaction on the VOCs associated with produced bread (23)
	Study of VOCs characterizing starter cultures isolated from alcoholic beverages and commercialized as flavoring agents for bakery products (19)
	Examination of VOCs associated with crumb and crust of breads prepared with gluten-free flours (27)
	Online monitoring of VOCs release in association with baking and toasting of breads from gluten-free flours (28)
Yoghurt	VOCs-based monitoring of fermentation in yoghurt using diverse commercial starter cultures (18)
Kefir	Evaluation of the impact on flavor of lactic acid bacteria strains to improve vitamin B2 content in kefir-like cereal-based beverages (29)
Cheese	Study of the impact of dairy system and individual animal factors on cheese volatiles (30)
Wine	Discernment of wines coming from diverse places and inoculated with different lactic acid bacteria (31)
	Evaluation of VOCs modulation after the inoculation of mixed starter cultures to promote alcoholic fermentation (24)
Beer	Monitoring the profile of hop-derived VOCs during the brewing process (32)

Monitoring Headspace VOCs during Yeasts Grown in a Standard Culture Media

Khomenko *et al.* (first experiment) investigated the online release of VOCs during the growth of diverse *S. cerevisiae* strains on agarized microbiological medium (*21*). To provide information about the screening potential of DIMS, this experiment simultaneously monitored six *S. cerevisiae* strains, each growing individually on the media over 11 days, with a measurement every 4 h, resulting in a considerable number of headspace analyses for each strain. In terms of chemical analysis, an average spectrum associated with each sampling point was found to consist of more than 300 peaks, of which a tentative identification was obtained for 70 of these (*21*). PCA of all six strains, including six replicates of each fermentation time point, and the uninoculated medium and empty vials explained 87.3% of the variance in the data over the first two principal components (Figure 1). This tailored data analysis of the resulting dataset allowed *i)* the study of the evolution of VOCs released during the growth (single strain volatilome) and *ii)* a preliminary evaluation of the intraspecies diversity of VOCs released within *S. cerevisiae* (*21*).

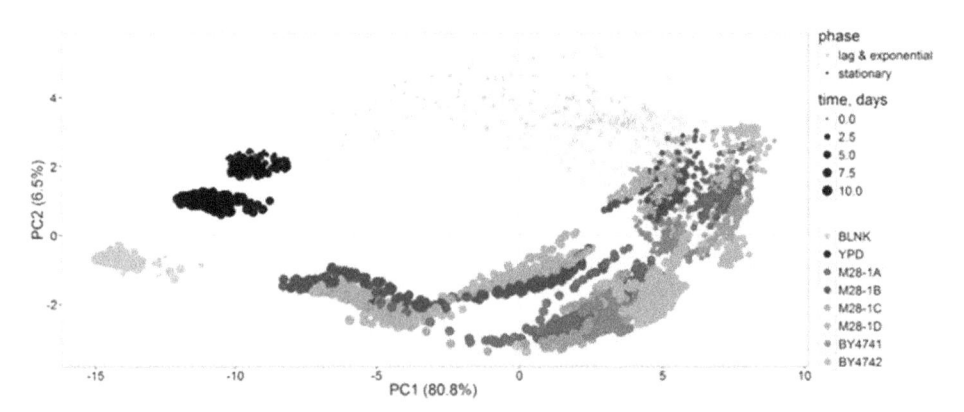

Figure 1. Score plot of PCA of VOCs associated with the different samples during an 11-day experiment. According to the legend, the diverse eight colors indicate the six yeast strains, the uninoculated medium and the empty vial (blank). The increase in the size of points designates the temporal progress of the experiments. Data are logarithmically transformed and centered before PCA. Reproduced with permission from reference (21). Copyright 2017 Springer Nature.

Based on this preliminary data exploration by unsupervised PCA, it was possible to select the mass peaks that made the largest contribution to the VOC evolution and those that allowed differentiation between yeasts. These peaks were separated and tentatively identified on the basis of their exact mass and simultaneous analysis using fastGC-PTR-TOF-MS (21). Summarizing, this experimental design, coupled with the chosen analytical and data analysis approach, allows the online high-throughput study of VOCs associated with the growth of different *S. cerevisiae* strains and enables the recognition of strain-specific behaviors (21). The extraction of these VOCs allowed the detection of novel strain-specific metabolic pathways. Finally, the technique demonstrated an excellent potential for phenotyping by allowing discrimination of genetically related strains, which is an innovative application of DIMS that has interesting applications in the field of basic and applied sciences (21). Chemical and/or biological changes are of crucial interest in the high-throughput screening of microbial strains for the selection of starter cultures that can be used to introduce novel flavor characteristics for the agro-food industry.

Study of VOC Release in Association with the Growth of Different Yeast Starter Cultures in a Real Food Matrix

Makhoul *et al*. (22) (second experiment) investigated the effect of four different starter cultures on VOC formation throughout bread production, specifically during the fermentation and after baking. Three starter cultures were from different *S. cerevisiae* strains (Y1, Y2 and Y3) and corresponding breads (B1, B2, and B3) were produced from the fermented doughs. Figure 2 shows the distinctive spectrum of *m/z* associated with the headspace of *i)* the dough during the fermentation and *ii)* the corresponding bread. Both of these mass spectra each consist of more than 400 separate peaks (22).

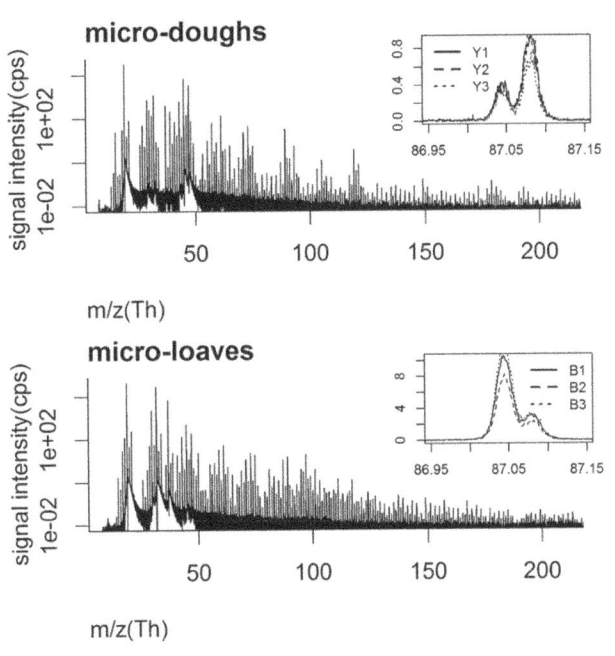

Figure 2. Average mass spectra of dough (top) and bread (bottom) (over a measurement time of 150–750 s). The insets on the top right report the peak profiles corresponding to the nominal mass m/z 87 for the trials Y1, Y2, and Y3 (top), and B1, B2, and B3 (bottom). Reproduced with permission from reference (22). Copyright 2014 John Wiley and Sons.

Following tailored calibration, the precise mass could be determined to three decimal places for each peak in the PTR-TOF-MS data. This allowed the estimation of the corresponding sum formula and, based on the scientific literature, it was possible to assign a tentative identification for a number of the sum formulas. The compounds tentatively identified fell into the following chemical classes: alkenes, esters, aldehydes, ketones, alcohols, carboxylic acids, sulfur compounds and some furan derivatives (22). A one-way ANOVA found 16 and 8 peaks that significantly ($p < 0.05$) changed as a function of the starter culture inoculated in the doughs and in the corresponding breads, respectively. The insets in Figure 2 show the trends for the nominal m/z 87 of the doughs inoculated with the three yeast and the corresponding breads (22). This nominal m/z 87 underlines the importance of achieving high mass resolution, which is made possible by the TOF mass analyzer, where m/z 87 was demonstrated to be the sum of masses m/z 87.044 (tentatively identified as diacetyl) and m/z 87.081 (tentatively identified as a C_5 aldehyde or ketone). The proportion of m/z 87.044 and m/z 87.081 differed between the dough and the resulting bread microloaves, with m/z 87.044 the dominant mass in the microloaves whereas m/z 87.081 was dominant in the doughs. Overall, the experimental approach using DIMS allowed the impact of the different inoculated microbial food cultures on the food volatilome to be categorized and understood in relation to their different potentials to contribute to bread flavor.

Testing the Effect on Flavor Formation of the Interaction between Yeast Starter Culture and Different Wheat Bread Flours

The third experiment increased experimental complexity further by using the potential of PTR-TOF-MS to measure not only differences among the starter cultures but also to evaluate the starter culture effects of different commercial microbial preparations on different wheat bread flours (23). In this experiment, doughs were prepared from four different flours (F1, durum wheat; F2, bread wheat type '00'; F3, bread wheat type '0'; or F4, Manitoba flour), then all doughs were inoculated with yeast starter culture 1 (Y1) or yeast starter culture 2 (Y2) (23). Analysis of the VOCs released during fermentation found that yeast type had a larger effect on VOC release than flour type (Figure 3) (23).

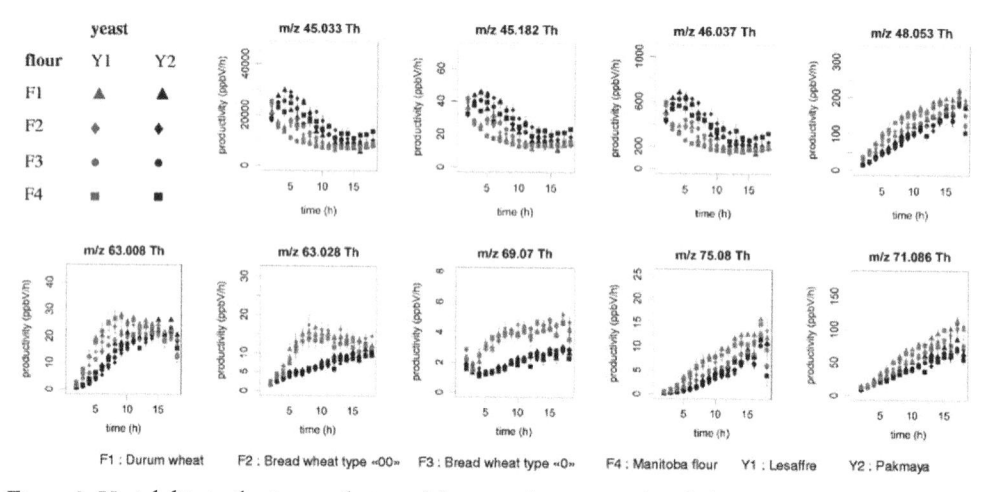

Figure 3. Variability in the time evolution of chosen m/z associated with dough fermentation. The two different scales of grey refer to the two yeast cultures investigated, while the different points indicate the four flour types. In general, each point denotes average productivity values. Y1 and Y2 are two different commercial yeast cultures. F1, durum wheat; F2, bread wheat type '00'; F3, bread wheat type '0'; F4, Manitoba flour. Reproduced with permission from reference (23). Copyright 2015 Springer Nature.

In the doughs, the yeast cultures exhibited a larger effect on VOC release over the fermentation time course than flour-type (Figure 3). However, a significant ($p<0.05$) interaction was found between yeast type and flour type, suggesting that VOC release does not only depend upon yeast type but also the flour type present (23). This interaction is more clearly displayed in Figure 4, where in addition to changes in peak m/z 47.014 (tentatively identified as formic acid) intensities with yeast type, changes in intensity can also be observed based on flour type (23). For formic acid, a clear synergistic interaction was found between Y1 and F3, considering this combination of experimental modes led to an amplified intensity compared with all the other possible combinations.

This study demonstrated the versatility and flexibility of PTR-TOF-MS by allowing the effect of a full factorial design of yeast type and flour type on mVOCs to be carried out in a reasonable time and in the process identifying potential advances in selecting starter cultures to achieve enhanced flavor for a given product/process (e.g., choice of a good yeast/flour combination).

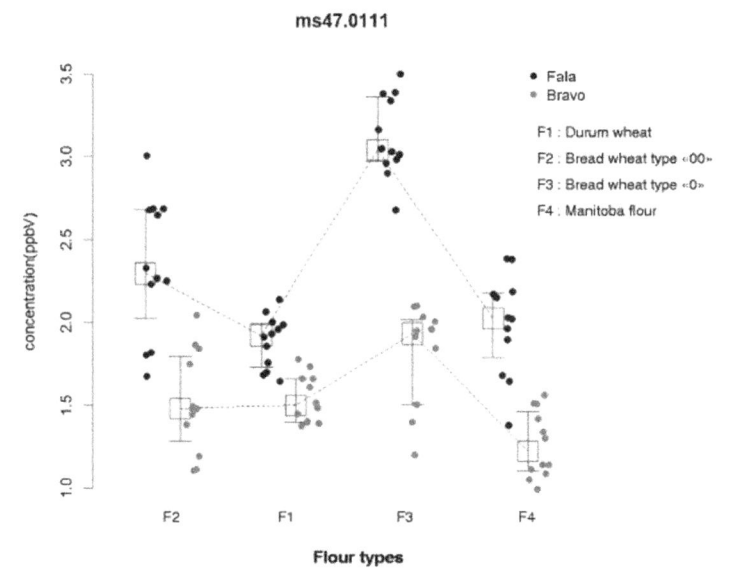

Figure 4. Variability of mass peak m/z 47.011 with the different combinations of yeast and flour. Reproduced with permission from reference (23). Copyright 2015 Springer Nature.

Volatilome Analysis to Estimate the Interaction between Different Combinations of Starter Cultures Grown in Multiple Food Matrices

In fermented food systems, not only biotic/abiotic interactions need to be considered, but also biotic/biotic interactions, in particular in the light of the increasing use of starter cultures consisting of multiple microbial species and/or strains within the same fermentation. In wine, for example, online measurements may be performed to improve the management of microbial resources during fermentation (33), and these measurements become more important when wishing to understand the effect of co-inoculation of non-*Saccharomyces* strains with *S. cerevisiae* on flavor generation (34). Berbegal *et al.* (24) (fourth experiment) used PTR-TOF-MS to investigate the impact on wine VOC release during fermentation from a full factorial combination of two *Saccharomyces* strains and two non-*Saccharomyces* strains in commercial grape juice (model matrix) and fresh grape must (real matrix), i.e., 4 yeast strains × 2 matrices. The two non-*Saccharomyces* strains belong to the *M. pulcherrima* and *T. delbrueckii* species. The PCA in Figure 5 summarizes the variability across the whole experimental plan, with the first two components explaining 84% of the total variability (24). Besides the general separation according to the two matrices (juice and must), different profiles in the distribution of four individual yeast and co-inoculated yeast in the PCA indicated *i)* the presence of a microbe/matrix effect and *ii)* the ability to monitor yeast species and strain interactions via VOC generation (24). The findings demonstrate why interest exists in DIMS techniques as a tool to understand the influence of the co-inoculation of diverse microbes on fermentation flavor development in a real food/beverage matrix. This ability to design appropriate combinations and/or multi-strain/species starter cultures in order to control and direct flavor generation of complex microbial food cultures is of outstanding interest.

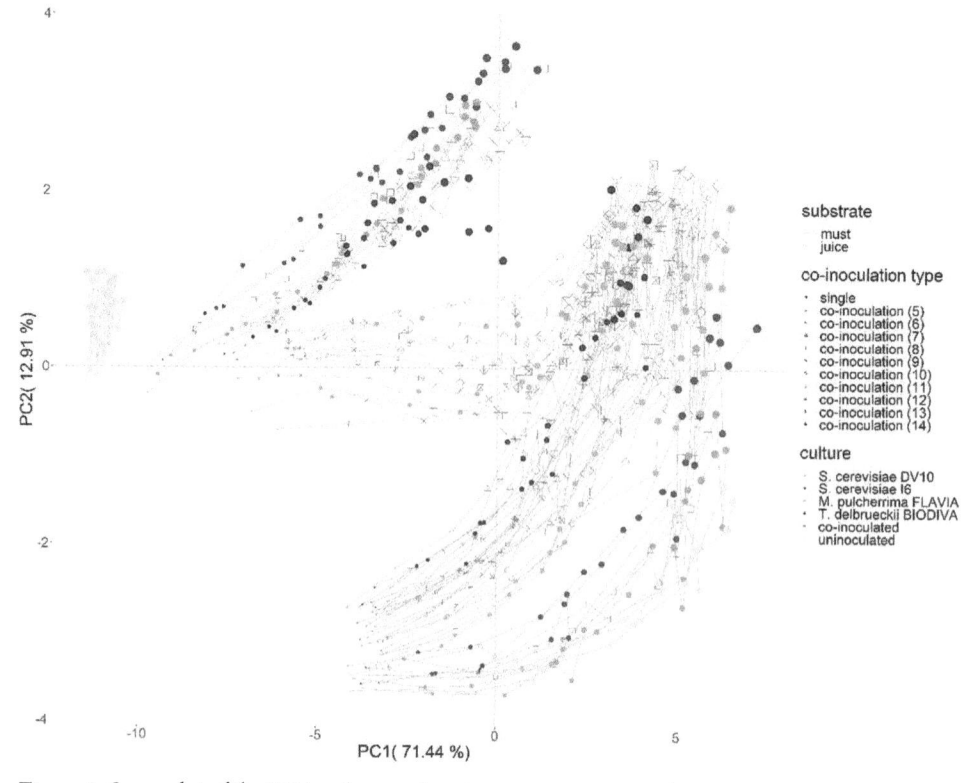

Figure 5. Score plot of the PCA, where each point represents a wine fermentation time point (point dimension increases proportionally with progress in fermentation time) from which VOCs were released during three days after yeast inoculation. Data were logarithmically transformed and centered. The following trials are summarized in the legend: (5) DV10, I6; (6) DV10, FLAVIA; (7) DV10, BIODIVA; (8) I6, FLAVIA; (9) I6, BIODIVA, (10) DV10, I6, FLAVIA; (11) DV10, I6, BIODIVA; (12) DV10, FLAVIA, BIODIVA; (13) I6, FLAVIA, BIODIVA; (14) DV10, I6, FLAVIA, BIODIVA. Reproduced with permission from reference (24). Copyright 2019 MDPI.

Investigating the Contribution of Headspace VOCs to Specific Flavor Attributes Produced by Different Starter Cultures in a Real Food Matrix

Finally, the analytical approach was applied to the investigation of commercial starter cultures with explicit flavor character claims. Capozzi *et al.* (*19*) (fifth experiment), used PTR-TOF-MS to uncover the potential VOCs behind the aroma marketing claims of three *S. cerevisiae* strains, isolated from alcoholic beverages (white wine, red wine, and beer) and commercialized in the bakery sector to introduce specific flavor characteristics in baked products. The flavor characters these cultures claim to introduce are *i)* 'a clean and delicate bouquet' (A16, strain isolated from white wine), *ii)* 'a fruity bouquet' (A17, strain isolated from red wine), and *iii)* 'a round and complex flavor contribution (characterized by buttery and exotic notes)' (A18, strain isolated from beer). The VOC analysis identified the presence of m/z 89.059, m/z 103.075 and m/z 117.093 that were tentatively identified as acetoin, ethyl butanoate/ethyl isobutanoate and ethyl propionate, respectively, and in concentrations that were indeed compatible with the aromatic characters stated for the three cultures evaluated (*19*) (Table 3) (when compared, as control (C), with a standard commercial starter culture without these particular aroma attributes).

Table 3. Concentrations for *m/z* 89.059, *m/z* 103.075 and *m/z* 117.093 corresponding to the cycles 10, 15, 20, and 24 (leavening process driven by *S. cerevisiae* bakery starter cultures C, A16, A17, and A18). Data reported from reference (*19*). Copyright 2016 MDPI.

m/z	Cycle	Concentrations (ppbv)			
		C	A16	A17	A18
89.060	10	173 ± 58 [b]	52 ± 3 [a]	157 ± 17 [b]	270 ± 37 [c]
89.060	15	314 ± 53 [b]	91 ± 14 [a]	478 ± 78 [c]	644 ± 88 [d]
89.060	20	396 ± 77 [b]	108 ± 15 [a]	625 ± 168 [c]	947 ± 137 [d]
89.060	24	661 ± 110 [b]	225 ± 20 [a]	1261 ± 102 [c]	1855 ± 182 [d]
103.075	10	1.7 ± 0.5 [b]	0.9 ± 0.2 [a]	1.8 ± 0.1 [b]	2.8 ± 0.3 [c]
103.075	15	2.2 ± 0.4 [b]	1.5 ± 0.3 [a]	3.0 ± 0.4 [c]	3.7 ± 0.4 [c]
103.075	20	2.4 ± 0.4 [a,b]	1.6 ± 0.4 [a]	3.0 ± 0.5 [b]	4.3 ± 0.6 [c]
103.075	24	3.2 ± 0.7 [a]	3.0 ± 0.7 [a]	5.1 ± 0.6 [b]	6.7 ± 0.6 [c]
117.093	10	2.1 ± 0.8 [b]	0.9 ± 0.1 [a]	2.7 ± 0.2 [b,c]	3.1 ± 0.4 [c]
117.093	15	3.0 ± 0.6 [b]	1.4 ± 0.2 [a]	6.0 ± 1.3 [c]	4.9 ± 0.4 [c]
117.093	20	3.3 ± 0.5 [a]	1.6 ± 0.3 [a]	6.1 ± 1.7 [b]	6.1 ± 0.9 [b]
117.093	24	4.9 ± 0.6 [a]	3.2 ± 0.3 [a]	11 ± 1 [b]	11 ± 1 [b]

Means with the same superscript letters are not significantly different (one-way ANOVA, $p<0.05$, Tukey test).

These results represent a starting point for high-throughput screenings of yeast biobanks where these particular *m/z* peaks (and their associated fragments) could be used as VOC markers to enable tailoring of flavor characters and identify flavor phenotypes (*19*) in relation to known flavor attributes. This would be supported by GC-MS analysis of selected samples to confirm peak identity.

The experiments described in this chapter are based on model microorganisms and model matrices that vary in complexity and include real food systems. In addition, the approaches used are straightforward to transfer to other microbes and other fermented foods/beverages and represent a new tool to foster innovation in the rapidly expanding field of fermented foods and food ingredients.

Conclusions

The importance of food fermentations for human health and wellbeing is the subject of continuous studies and discoveries (*10, 35–37*). The issue is of particular interest if one considers that fermented foods and beverages account for more than 30% of global food consumption (*9*). The possibility of modulating sensory quality is crucial because it influences consumer choice and may be used to promote consumption of several fermented products that are well recognized for their nutritional and functional features (*10, 35, 38, 39*). In this context, mVOCs during fermentation are of great importance due to their crucial influence on flavor and consumer acceptability (*40*). In addition, the diverse VOCs associated with a given system, the volatilome (*21, 42*), can be used to monitor and control bioprocesses and, more generally, to assess the quality and safety of foods (*4, 8, 41*).

Globally, fermentation is considered a key technology when considering sustainable food production as it contributes to reduced resource dissipation, reduced energy use, improved food quality and safety, preservation of microbial diversity, creation of value for food by-products, and reduction of food waste (*43–46*). In addition, remaining in the field of 'green' applications, manipulation of fermentation conditions and starter culture can be used to reduce label complexity ('clean food label') and mitigate the negative impacts of climate changes on food quality (*44, 47*). Considering *i)* that the number of fermented foods is estimated to be in the thousands of products, *ii)* that an extensive variety of microbial species are involved, and *iii)* the importance of monitoring for bioprocess optimization, it is evident of the attentiveness for analytical techniques that allow online monitoring and high-throughput screening based on fingerprinting of flavor compounds (*11, 18, 35*).

DIMS technologies represent an interesting and versatile set of analytical instruments that are well-tailored to the analysis of the volatilome associated with food fermentations, such as following the inoculation of starter cultures and the impact on flavor-active VOCs (*4, 18, 40*). The non-destructive nature and ability to avoid sample pre-treatments (including extraction) makes DIMS a desirable tool for low-cost and time-saving analytical designs, low-impact analytical solutions, real-time monitoring and high-throughput screening (*1, 4*). Typically, analytical approaches used for VOC analysis exploit instrumental strategies which either optimize sample throughput or the chemical information obtained. PTR-TOF-MS offers a valuable compromise, providing rapid and direct measurements together with an output containing highly valuable chemical information. Other benefits of PTR-TOF-MS use are the possibility *i)* of rapid and continuous measurement, *ii)* to use intense and pure sources of precursor ions, and *iii)* to control ionization conditions (pressure, temperature and drift voltage). These characteristics allow versatile applications while maintaining a high degree of standardization. Exciting opportunities exist for experiments investigating all aspects of screening, design and application of flavor generating starter cultures for the vast and heterogeneous sector of fermented foods and beverages.

The ability to select tailored green analytical techniques (e.g., DIMS methodologies) to sustain the development of green biotechnological solutions for industry (e.g., starter culture technology) represents an intriguing strategy to aid a transformation to a sustainable agro-food system (*43, 48*). This is an aspect of particular interest to achieve a holistic effort towards sustainable development, not only from the farm to the fork, but also from the research through to development and up to production.

Finally, it is important to underline the opportunities in exploiting the untargeted nature of DIMS technologies (*49, 50*). This analytical feature is of huge significance, merging basic and applied science to provide an improved opportunity to produce both original solutions and advances of knowledge at the interface between agro-food chemistry and microbiological sciences.

Acknowledgments

We would like to thank all the co-authors of the scientific works described in this chapter and Massimo Franchi of the Institute of Sciences of Food Production—CNR for the skilled technical support provided during the realization of this work.

Notes

All authors equally contributed to this work.

References

1. Biasioli, F.; Yeretzian, C.; Märk, T. D.; Dewulf, J.; Van Langenhove, H. Direct-Injection Mass Spectrometry Adds the Time Dimension to (B)VOC Analysis. *TrAC Trends in Analytical Chemistry* **2011**, *30* (7), 1003–1017.
2. Xu, X.; Li, W.; Li, T.; Zhang, K.; Song, Q.; Liu, L.; Tu, P.; Wang, Y.; Song, Y.; Li, J. Direct Infusion-Three-Dimensional-Mass Spectrometry Enables Rapid Chemome Comparison among Herbal Medicines. *Anal Chem* **2020**, *92* (11), 7646–7656.
3. Romano, A.; Capozzi, V.; Spano, G.; Biasioli, F. Proton Transfer Reaction–Mass Spectrometry: Online and Rapid Determination of Volatile Organic Compounds of Microbial Origin. *Appl Microbiol Biotechnol* **2015**, *99* (9), 3787–3795.
4. Capozzi, V.; Yener, S.; Khomenko, I.; Farneti, B.; Cappellin, L.; Gasperi, F.; Scampicchio, M.; Biasioli, F. PTR-ToF-MS Coupled with an Automated Sampling System and Tailored Data Analysis for Food Studies: Bioprocess Monitoring, Screening and Nose-Space Analysis. *J Vis Exp* **2017** (No. 123), e54075.
5. Gliszczyńska-Świgło, A.; Chmielewski, J. Electronic Nose as a Tool for Monitoring the Authenticity of Food. A Review. *Food Anal. Methods* **2017**, *10* (6), 1800–1816.
6. Carrapiso, A. I.; Noseda, B.; García, C.; Reina, R.; Sánchez del Pulgar, J.; Devlieghere, F. SIFT-MS Analysis of Iberian Hams from Pigs Reared under Different Conditions. *Meat Science* **2015**, *104*, 8–13.
7. Capozzi, V.; Lonzarich, V.; Khomenko, I.; Cappellin, L.; Navarini, L.; Biasioli, F. Unveiling the Molecular Basis of Mascarpone Cheese Aroma: VOCs Analysis by SPME-GC/MS and PTR-ToF-MS. *Molecules* **2020**, *25* (5), 1242.
8. Makhoul, S.; Yener, S.; Khomenko, I.; Capozzi, V.; Cappellin, L.; Aprea, E.; Scampicchio, M.; Gasperi, F.; Biasioli, F. Rapid Non-Invasive Quality Control of Semi-Finished Products for the Food Industry by Direct Injection Mass Spectrometry Headspace Analysis: The Case of Milk Powder, Whey Powder and Anhydrous Milk Fat. *Journal of Mass Spectrometry* **2016**, *51* (9), 782–791.
9. Vogel, R. F.; Hammes, W. P.; Habermeyer, M.; Engel, K.-H.; Knorr, D.; Eisenbrand, G. Microbial Food Cultures – Opinion of the Senate Commission on Food Safety (SKLM) of the German Research Foundation (DFG). *Molecular Nutrition & Food Research* **2011**, *55* (4), 654–662.
10. Marco, M. L.; Heeney, D.; Binda, S.; Cifelli, C. J.; Cotter, P. D.; Foligné, B.; Gänzle, M.; Kort, R.; Pasin, G.; Pihlanto, A.; Smid, E. J.; Hutkins, R. Health Benefits of Fermented Foods: Microbiota and Beyond. *Curr. Opin. Biotechnol.* **2017**, *44*, 94–102.
11. Tamang, J. P.; Watanabe, K.; Holzapfel, W. H. Review: Diversity of Microorganisms in Global Fermented Foods and Beverages. *Front. Microbiol.* **2016**, *7*.
12. Capozzi, V.; Fragasso, M.; Romaniello, R.; Berbegal, C.; Russo, P.; Spano, G. Spontaneous Food Fermentations and Potential Risks for Human Health. *Fermentation* **2017**, *3* (4), 49.
13. Leroy, F.; De Vuyst, L. Lactic Acid Bacteria as Functional Starter Cultures for the Food Fermentation Industry. *Trends in Food Science & Technology* **2004**, *15* (2), 67–78.
14. Mozzi, F.; Ortiz, M. E.; Bleckwedel, J.; De Vuyst, L.; Pescuma, M. Metabolomics as a Tool for the Comprehensive Understanding of Fermented and Functional Foods with Lactic Acid Bacteria. *Food Research International* **2013**, *54* (1), 1152–1161.

15. Chen, G.; Chen, C.; Lei, Z. Meta-Omics Insights in the Microbial Community Profiling and Functional Characterization of Fermented Foods. *Trends in Food Science & Technology* **2017**, *65*, 23–31.
16. Liu, H.; Sun, B. Effect of Fermentation Processing on the Flavor of Baijiu. *J. Agric. Food Chem.* **2018**, *66* (22), 5425–5432. https://doi.org/10.1021/acs.jafc.8b00692.
17. Zhang, Y.; Fraatz, M. A.; Horlamus, F.; Quitmann, H.; Zorn, H. Identification of Potent Odorants in a Novel Nonalcoholic Beverage Produced by Fermentation of Wort with Shiitake (Lentinula edodes). *J. Agric. Food Chem.* **2014**, *62* (18), 4195–4203.
18. Benozzi, E.; Romano, A.; Capozzi, V.; Makhoul, S.; Cappellin, L.; Khomenko, I.; Aprea, E.; Scampicchio, M.; Spano, G.; Märk, T. D.; Gasperi, F.; Biasioli, F. Monitoring of Lactic Fermentation Driven by Different Starter Cultures via Direct Injection Mass Spectrometric Analysis of Flavour-Related Volatile Compounds. *Food Research International* **2015**, *76*, 682–688.
19. Capozzi, V.; Makhoul, S.; Aprea, E.; Romano, A.; Cappellin, L.; Sanchez Jimena, A.; Spano, G.; Gasperi, F.; Scampicchio, M.; Biasioli, F. PTR-MS Characterization of VOCs Associated with Commercial Aromatic Bakery Yeasts of Wine and Beer Origin. *Molecules* **2016**, *21* (4), 483.
20. Van Kerrebroeck, S.; Vercammen, J.; Wuyts, R.; De Vuyst, L. Selected Ion Flow Tube–Mass Spectrometry for Online Monitoring of Submerged Fermentations: A Case Study of Sourdough Fermentation. *J. Agric. Food Chem.* **2015**, *63* (3), 829–835.
21. Khomenko, I.; Stefanini, I.; Cappellin, L.; Cappelletti, V.; Franceschi, P.; Cavalieri, D.; Märk, T. D.; Biasioli, F. Non-Invasive Real Time Monitoring of Yeast Volatilome by PTR-ToF-MS. *Metabolomics* **2017**, *13* (10), 118.
22. Makhoul, S.; Romano, A.; Cappellin, L.; Spano, G.; Capozzi, V.; Benozzi, E.; Märk, T. D.; Aprea, E.; Gasperi, F.; El-Nakat, H.; Guzzo, J.; Biasioli, F. Proton-Transfer-Reaction Mass Spectrometry for the Study of the Production of Volatile Compounds by Bakery Yeast Starters. *Journal of Mass Spectrometry* **2014**, *49* (9), 850–859.
23. Makhoul, S.; Romano, A.; Capozzi, V.; Spano, G.; Aprea, E.; Cappellin, L.; Benozzi, E.; Scampicchio, M.; Märk, T. D.; Gasperi, F.; El-Nakat, H.; Guzzo, J.; Biasioli, F. Volatile Compound Production During the Bread-Making Process: Effect of Flour, Yeast and Their Interaction. *Food Bioprocess Technol* **2015**, *8* (9), 1925–1937.
24. Berbegal, C.; Khomenko, I.; Russo, P.; Spano, G.; Fragasso, M.; Biasioli, F.; Capozzi, V. PTR-ToF-MS for the Online Monitoring of Alcoholic Fermentation in Wine: Assessment of VOCs Variability Associated with Different Combinations of *Saccharomyces*/Non-*Saccharomyces* as a Case-Study. *Fermentation* **2020**, *6* (2), 55.
25. Romano, A.; Fischer, L.; Herbig, J.; Campbell-Sills, H.; Coulon, J.; Lucas, P.; Cappellin, L.; Biasioli, F. Wine Analysis by FastGC Proton-Transfer Reaction-Time-of-Flight-Mass Spectrometry. *International Journal of Mass Spectrometry* **2014**, *369*, 81–86.
26. Acierno, V.; Fasciani, G.; Kiani, S.; Caligiani, A.; van Ruth, S. PTR-QiToF-MS and HSI for the Characterization of Fermented Cocoa Beans from Different Origins. *Food Chemistry* **2019**, *289*, 591–602.
27. Pico, J.; Khomenko, I.; Capozzi, V.; Navarini, L.; Bernal, J.; Gómez, M.; Biasioli, F. Analysis of volatile organic compounds in crumb and crust of different baked and toasted gluten-free

breads by direct PTR-ToF-MS and fast-GC-PTR-ToF-MS. *Journal of Mass Spectrometry* **2018**, *53* (9), 893–902.

28. Pico, J.; Khomenko, I.; Capozzi, V.; Navarini, L.; Biasioli, F. Real-Time Monitoring of Volatile Compounds Losses in the Oven during Baking and Toasting of Gluten-Free Bread Doughs: A PTR-MS Evidence. *Foods* **2020**, *9* (10), 1498.

29. Yépez, A.; Russo, P.; Spano, G.; Khomenko, I.; Biasioli, F.; Capozzi, V.; Aznar, R. In Situ Riboflavin Fortification of Different Kefir-like Cereal-Based Beverages Using Selected Andean LAB Strains. *Food Microbiology* **2019**, *77*, 61–68.

30. Bergamaschi, M.; Biasioli, F.; Cappellin, L.; Cecchinato, A.; Cipolat-Gotet, C.; Cornu, A.; Gasperi, F.; Martin, B.; Bittante, G. Proton Transfer Reaction Time-of-Flight Mass Spectrometry: A high-throughput and innovative method to study the influence of dairy system and cow characteristics on the volatile compound fingerprint of cheeses. *Journal of dairy science* **2015**, *98* (12), 8414–8427.

31. Campbell-Sills, H.; Capozzi, V.; Romano, A.; Cappellin, L.; Spano, G.; Breniaux, M.; Lucas, P.; Biasioli, F. Advances in Wine Analysis by PTR-ToF-MS: Optimization of the Method and Discrimination of Wines from Different Geographical Origins and Fermented with Different Malolactic Starters. *International Journal of Mass Spectrometry* **2016**, *397–398*, 42–51.

32. Richter, T. M.; Silcock, P.; Algarra, A.; Eyres, G. T.; Capozzi, V.; Bremer, P. J.; Biasioli, F. Evaluation of PTR-ToF-MS as a Tool to Track the Behavior of Hop-Derived Compounds during the Fermentation of Beer. *Food Research International* **2018**, *111*, 582–589.

33. Silva Ferreira, A. C.; Monforte, A. R.; Teixeira, C. S.; Martins, R.; Fairbairn, S.; Bauer, F. F. Monitoring Alcoholic Fermentation: An Untargeted Approach. *J. Agric. Food Chem.* **2014**, *62* (28), 6784–6793.

34. Roudil, L.; Russo, P.; Berbegal, C.; Albertin, W.; Spano, G.; Capozzi, V. Non-Saccharomyces Commercial Starter Cultures: Scientific Trends, Recent Patents and Innovation in the Wine Sector. *Recent Patents on Food, Nutrition & Agriculture* **2020**, *11* (1), 27–39. 3.

35. Dimidi, E.; Cox, S. R.; Rossi, M.; Whelan, K. Fermented Foods: Definitions and Characteristics, Impact on the Gut Microbiota and Effects on Gastrointestinal Health and Disease. *Nutrients* **2019**, *11* (8), 1806.

36. Gao, J.; Mao, K.; Wang, X.; Mi, S.; Fu, M.; Li, X.; Xiao, J.; Simal-Gandara, J.; Sang, Y. Tibet Kefir Milk Regulated Metabolic Changes Induced by High-Fat Diet via Amino Acids, Bile Acids, and Equol Metabolism in Human-Microbiota-Associated Rats. *J. Agric. Food Chem.* **2021**, *69* (23), 6720–6732.

37. Que, Z.; Ma, T.; Shang, Y.; Ge, Q.; Zhang, Q.; Xu, P.; Zhang, J.; Francoise, U.; Liu, X.; Sun, X. Microorganisms: Producers of Melatonin in Fermented Foods and Beverages. *J. Agric. Food Chem.* **2020**, *68* (17), 4799–4811.

38. Zhou, Q.; Xue, B.; Gu, R.; Li, P.; Gu, Q. *Lactobacillus plantarum* ZJ316 Attenuates *Helicobacter pylori*-Induced Gastritis in C57BL/6 Mice. *J. Agric. Food Chem.* **2021**, *69* (23), 6510–6523.

39. Utz, F.; Kreissl, J.; Stark, T. D.; Schmid, C.; Tanger, C.; Kulozik, U.; Hofmann, T.; Dawid, C. Sensomics-Assisted Flavor Decoding of Dairy Model Systems and Flavor Reconstitution Experiments. *J. Agric. Food Chem.* **2021**, *69* (23), 6588–6600.

40. Spence, C. What Is the Relationship between the Presence of Volatile Organic Compounds in Food and Drink Products and Multisensory Flavour Perception. *Foods* **2021**, *10* (7), 1570.

41. Bahroun, N. H. O.; Perry, J. D.; Stanforth, S. P.; Dean, J. R. Use of Exogenous Volatile Organic Compounds to Detect Salmonella in Milk. *Analytica Chimica Acta* **2018**, *1028*, 121–130.
42. Alves, M. L.; Bento-Silva, A.; Gaspar, D.; Paulo, M.; Brites, C.; Mendes-Moreira, P.; Bronze, M. do R.; Malosetti, M.; van Eeuwijk, F.; Vaz Patto, M. C. Volatilome–Genome-Wide Association Study on Wholemeal Maize Flour. *J. Agric. Food Chem.* **2020**, *68* (29), 7809–7818.
43. Timmis, K.; de Vos, W. M.; Ramos, J. L.; Vlaeminck, S. E.; Prieto, A.; Danchin, A.; Verstraete, W.; Lorenzo, V. de; Lee, S. Y.; Brüssow, H.; Timmis, J. K.; Singh, B. K. The Contribution of Microbial Biotechnology to Sustainable Development Goals. *Microbial Biotechnology* **2017**, *10* (5), 984–987.
44. Capozzi, V.; Fragasso, M.; Bimbo, F. Microbial Resources, Fermentation and Reduction of Negative Externalities in Food Systems: Patterns toward Sustainability and Resilience. *Fermentation* **2021**, *7* (2), 54.
45. Capozzi, V.; Fragasso, M.; Russo, P. Microbiological Safety and the Management of Microbial Resources in Artisanal Foods and Beverages: The Need for a Transdisciplinary Assessment to Conciliate Actual Trends and Risks Avoidance. *Microorganisms* **2020**, *8* (2), 306.
46. De Vero, L.; Boniotti, M. B.; Budroni, M.; Buzzini, P.; Cassanelli, S.; Comunian, R.; Gullo, M.; Logrieco, A. F.; Mannazzu, I.; Musumeci, R.; Perugini, I.; Perrone, G.; Pulvirenti, A.; Romano, P.; Turchetti, B.; Varese, G. C. Preservation, Characterization and Exploitation of Microbial Biodiversity: The Perspective of the Italian Network of Culture Collections. *Microorganisms* **2019**, *7* (12), 685.
47. Berbegal, C.; Fragasso, M.; Russo, P.; Bimbo, F.; Grieco, F.; Spano, G.; Capozzi, V. Climate Changes and Food Quality: The Potential of Microbial Activities as Mitigating Strategies in the Wine Sector. *Fermentation* **2019**, *5* (4), 85.
48. Ganesh, K. N.; Zhang, D.; Miller, S. J.; Rossen, K.; Chirik, P. J.; Kozlowski, M. C.; Zimmerman, J. B.; Brooks, B. W.; Savage, P. E.; Allen, D. T.; Voutchkova-Kostal, A. M. Green Chemistry: A Framework for a Sustainable Future. *Environ. Sci. Technol.* **2021**, *55* (13), 8459–8463.
49. Clark, T. N.; Houriet, J.; Vidar, W. S.; Kellogg, J. J.; Todd, D. A.; Cech, N. B.; Linington, R. G. Interlaboratory Comparison of Untargeted Mass Spectrometry Data Uncovers Underlying Causes for Variability. *J. Nat. Prod.* **2021**, *84* (3), 824–835.
50. Kirwan, J. A.; Broadhurst, D. I.; Davidson, R. L.; Viant, M. R. Characterising and Correcting Batch Variation in an Automated Direct Infusion Mass Spectrometry (DIMS) Metabolomics Workflow. *Anal Bioanal Chem* **2013**, *405* (15), 5147–5157.

Chapter 11

Monitoring of Acrolein, Acetaldehyde and 1,3-Butadiene in Fumes Emitted during Deep-Frying of Potato Pieces in Rapeseed Oil Using PTR-MS

Wiktoria Wichrowska and Tomasz Majchrzak[*]

Department of Analytical Chemistry, Faculty of Chemistry, Gdańsk University of Technology, Gabriela Narutowicza 11/12, 80-039 Gdańsk, Poland
[*]Email: tomasz.majchrzak@pg.edu.pl

The process of frying foods in oil leads to the production and release of many volatile organic compounds (VOCs), including those with toxic and carcinogenic properties. Proton transfer reaction-mass spectrometry (PTR-MS) was used to monitor the concentration of these compounds in real-time. This chapter reports on experiments that show how the concentrations of selected, toxic volatiles, namely acrolein, acetaldehyde, and 1,3-butadiene, change during the early stage of frying in rapeseed oil, as well as over long-term frying of 3 h. The concentration of acrolein in frying fumes was the highest among these compounds, exceeding 3 ppm$_v$ during long-term frying. The studies on emissions in the early stage of frying involved changing selected parameters, such as the period of oil heating, food immersion, frying, and the over-frying period. Concentrations of the targeted compounds in the frying fumes were observed to increase with frying time. Based on these data, this chapter discusses how parameters, such as oil temperature, food type, and the amount of food, affect the release kinetics of these volatiles. The results indicate that the key factor influencing the pattern of emission of these compounds is water present in the food, which is responsible for the initial release of volatiles caused by the sudden evaporation of water. Considering the cumulative release curves it can be observed that for acrolein, acetaldehyde, and 1,3-butadiene the emission profiles can be described by a second degree polynomial with a determination coefficient exceeding $R^2 = 0.999$.

Direct Injection Mass Spectrometry as a Tool for Characterizing Volatiles in Frying Fumes

In the past decade, direct injection mass spectrometry (DIMS) methods have been gaining popularity in food analysis. This is evidenced by the number of published research articles and

© 2021 American Chemical Society

review papers (*1–3*). The use of DIMS in the analysis of volatile organic compounds (VOCs) emitted from food represents a valuable approach in assessing the freshness of food products (*4–6*), their authenticity (*7–9*), or their quality, also in terms of aroma (*10–12*). Thus, DIMS techniques are often used in the testing of raw materials and the final products. These analytical techniques are unique since they enable real-time profiling of volatiles, which makes them a useful tool in the investigations of changes occurring during food processing.

To date, DIMS has been used in monitoring diverse processes, including cooking (*13, 14*), roasting (*15*), baking and toasting (*16*), as well as frying (*17*), among others. Heat treatment plays an important role in each of these processes. Increased temperature serves to obtain more easily digestible food, increase microbiological safety, or to gain appropriate food characteristics, such as taste, aroma, color, or texture. During these processes, a plethora of chemical reactions occur, including oxidation (*18*), hydrolysis (*19*), dehydration (*20*), Maillard reaction (*21*), and caramelization (*22*), to name but a few. Each of these reactions results in the formation of a variety of VOCs, which in turn can provide valuable analytical information. For example, changes in the concentration of 1-heptanol in the fumes emitted during frying with rapeseed oil may indicate a loss of food quality (*23*). Furthermore, products of thermal food processing might be toxic, hence measuring their concentrations is an important endeavor.

During frying, both the foodstuffs and the oil used for frying are transformed. The main contribution of the oil is to facilitate heat transfer, although its role is also to give adequate flavor to the food. The main reaction that occurs in heated oil is the oxidation of triacylglycerols (*24*). This reaction, as well as subsequent secondary reactions, results mostly in the emission of short-chain saturated and unsaturated aldehydes, alcohols, and aromatics. Understanding the kinetics of the release of these compounds may give an insight into this extremely complex process of frying.

For the past five years, experiments conducted at the Department of Analytical Chemistry of Gdańsk University of Technology, Poland, have focused on characterizing the volatiles emitted during deep-frying. This chapter contains a description of a part of a study in which a proton transfer reaction-mass spectrometer (PTR-MS) was used to track changes in the composition of fumes emitted during frying.

Toxic Volatile Products of Frying

During frying, a myriad of chemical compounds are formed as a result of a chain of reactions, including lipid oxidation (paramount reaction), thermal food processing, or secondary reactions at elevated temperatures. Oxidation is a three-step radical reaction. The first step, initiation, is the slowest and determines the rate of the entire process and defines the oxidative stability of the oil. At its core is the detachment of a hydrogen atom from an unsaturated fatty acid (USFA) to form an alkyl radical. In the next stage, propagation, the free alkyl radical reacts with oxygen to form a peroxy radical. Further reaction with USFA leads to hydroperoxide. Hydroperoxides generated at this stage are chemically unstable compounds with O-O bonds and are subject to further transformations. The interaction of two radicals results in termination and thus, as the reaction chain stops, non-radical molecules that are not able to continue the reaction are formed. The products formed at this stage are termed "secondary oxidation products" (SOPs) (*20, 24, 25*).

Volatiles emitted during frying contribute to the aroma and flavor of food and can be directly linked to the quality of the oil. High concentrations of compounds in oil fumes can, especially in the case of prolonged exposure, negatively affect health. Products such as carbonyl compounds, benzene, 1,3-butadiene, as well as semi-volatile polyaromatic hyrdocarbons (PAHs) are considered toxic by-products of food frying (26–28). Research conducted at Gdańsk has mainly focused on the chemical compounds that could be indicators of the oil quality, paying specific attention to the analysis of saturated and unsaturated volatile aldehydes, which are indicative of lipid oxidation (29, 30). Among these aldehydes, acrolein, which is the simplest α,β-unsaturated aldehyde, is one of the most common. Acrolein is reported to have toxic properties, for instance, it is classified as probably carcinogenic (IARC Group 2A).

Acrolein is a relatively reactive compound, with a half-life of 0.5-1.2 days (31). Extensive studies of its toxicity have been made, including research on lung cancer cases among Chinese women in relation to cooking (26). The conclusion from this work was that cooking fumes, and specifically acrolein present therein, can be considered carcinogenic, especially after long-term exposure. Further, this compound can induce serious mutagenic effects and represents a significant inhalation hazard in chronic obstructive pulmonary disease (COPD) (31). Moreover, reports indicate that acrolein passes through the alveolar capillary membrane and thus may contribute to cardiovascular disease (32) and excerbate asthma in children (33). The main sources of acrolein that directly affects human health are cigarette smoke and the frying process during food preparation. The mechanism of acrolein formation during frying is not clearly defined. The most probable mechanism assumes an important role of glycerol as a precursor of acrolein formation, although formation from polyunsaturated fatty acids (PUFAs) by lipid hydroperoxidation is also likely (34).

Apart from acrolein, other compounds that may have toxic effects were dected in frying fumes. Figure 1 depicts selected ions, detected by PTR-MS, that may correspond to the emission of compounds considered to be carcinogenic (IARC Group 1), namely 1,3-butadiene (m/z 55.054 – from hereon referred to as the m55 peak), probably carcinogenic, the aforementioned acrolein (m/z 57.033 – m57), or the possibly carcinogenic (IARC Group 2B) acetaldehyde (m/z 45.033 – m45). The abundance of each of these compounds were observed to increase as a result of food frying. Acetaldehyde is one of the simplest aldehydes, yet one of the most reactive. In the body, it can manifest carcinogenic properties, particularly with regards to gastric cancer (35). Although its toxicity is primarily linked with alcohol consumption, acetaldehyde is also a major component of oil fumes (28). In turn, 1,3-butadiene is also found in frying fumes (36), and its presence may have important implications for cancer risk (26, 37). Its metabolite, 3,4-epoxybutane-1,2-diol (EBD), is likely responsible for its carcinogenicity (38).

It is important to be aware that as long as there is no hard evidence that the emitted amounts of these compounds are harmful to human health, it is misleading to draw direct conclusions on their toxicological effects, especially since the experiments presented here were conducted in conditions differing from those commonly found in kitchens. The conducted experiments took place under controlled conditions, in a closed vessel, and with forced air flow. By characterizing the release profile of these chemical compounds, however, it is possible to demonstrate how they are emitted during frying, from the moment when food is placed in heated oil to the retrieval of the cooked product from the oil.

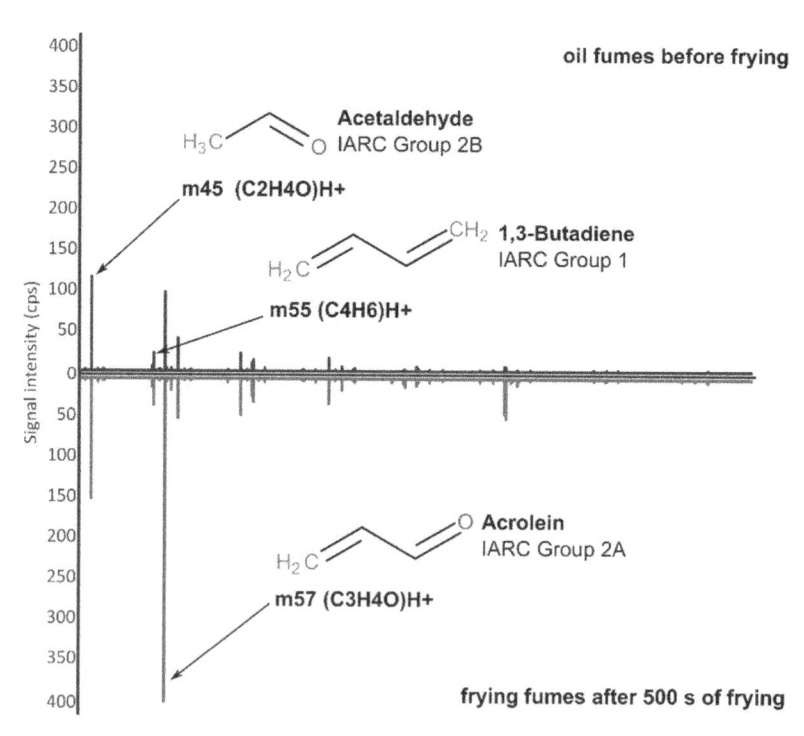

Figure 1. PTR-MS mass spectral fingerprints of fumes composition immediately before placing a potato cube into rapeseed oil heated to 180°C (top) and after 500 s of frying (bottom). The selected m/z values represent the selected toxic volatile thermal degradation products, namely acetaldehyde (m/z 45.033 – m45), 1,3-butadiene (m/z 55.054 – m55), and acrolein (m/z 57.033 – m57).

Monitoring Acetaldehyde, 1,3-Butadiene and Acrolein in Frying Fumes by PTR-MS

Two experiments are presented in this chapter, for which the experimental details and part of the results have been published elsewhere (17, 23). The experimental details are nevertheless summarized here. Frying fumes were collected in real-time using two sampling setups, depicted in Figure 2. A PTR-TOF 1000 ultra instrument (Ionicon Analytik GmbH, Innsbruck, Austria) was used in both experiments. The PTR-MS equipped with a time-of-flight (TOF) analyzer allows real-time monitoring of multiple VOCs simultaneously, and because of the high resolution, multiple isobaric compounds can be distinguished from each other. An overview of this technique can be found in Chatper 1, and more comprehensively in published scholarly works (39–41). The instrument used in this study exhibited a resolution of approximately R=1500, with a sensitivity at the single-digit ppb level. Spectra were registered every second of the measurements. The temperatures of the drift tube and the transfer line were both set to 70°C and the measurements were performed in standard mode (the PTR-TOF 1000 ultra enables measuring in funnel mode, which was not performed in these studies). In both cases, PTR-MS Viever v.3.4.2.1 (Ionicon Analytik GmbH, Innsbruck, Austria) was used for data analysis. All reported concentrations are expressed as calculated mixing ratios using the standard reaction rate coefficient $k = 2.0 \times 10^{-9}$ cm^{3}s^{-1}.

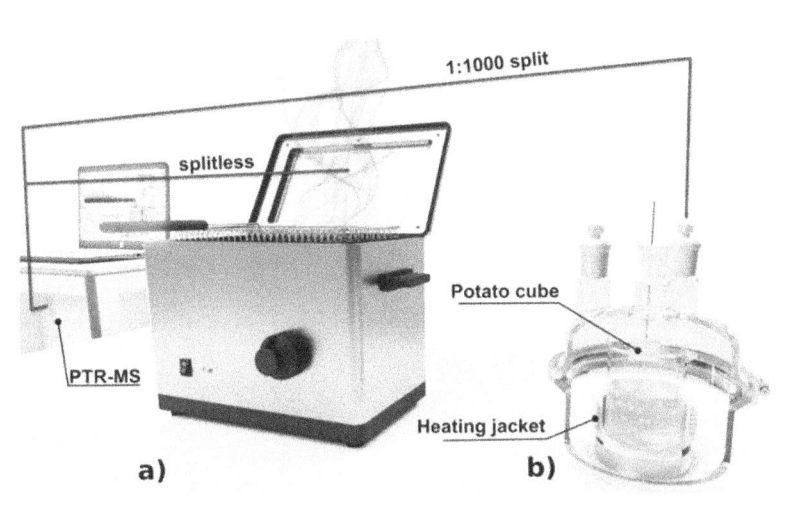

Figure 2. Two sample delivery methods for PTR-MS analysis: a) a deep-fryer placed under a fume hood, used in a long frying (3 h) experiment. The PTR-MS inlet was placed ~50 cm above the oil surface in order to prevent the capillary from clogging with oil condensate, and b) an in-house frying fumes delivery system for early-stage VOC monitoring. The oil (50 mL) was placed in a glass beaker and the frying temperature was controlled by a proportional–integral–derivative (PID) controller. The potato cube was immersed in the heated oil by lowering the metal hook. Both the potato cube and the oil were enclosed in the glass reactor. The fumes were carried by a clean air stream (500 sccm). Before entering the heated transfer line (70°C) the fumes passed a fabric filter for the oil and semi-volatile compounds to condensate. The fumes were then split using a T-peace and 1 sccm was diluted (1:1000) and directed into the PTR-MS.

Volatile Emissions during Prolonged Frying

The first experiment focused on the investigation of the composition of fumes during prolonged frying. In this experiment, pre-made (industrial) potato fries were fried in rapeseed oil (2 L) heated to ~160°C in a deep-fryer. Each portion of fries had a mass of 100 g and was fried for ca. 10 min until the desired doneness (the fries were ready for eating). The E/N of the PTR-MS was set to a value of 120 Td and the undiluted sample was introduced into the system with a flow of 100 sccm. The inlet capillary was placed ca. 50 cm above the oil surface to prevent the capillary from clogging with oil mist.

The experiment involved real-time measurements of VOC concentrations during the frying of potato fries in rapeseed oil. Each batch of fries was replaced approximately every 10 min, and the experiment itself lasted 3 h. As the frying progressed, a significant increase in the peroxide value was observed, from 2.69 mEqO$_2$ kg^{-1} in the case of fresh oil, to 17.66 mEqO2 kg^{-1} after 3 h. Focusing on the three potentially toxic candidates, it can be seen that the emissions of these compounds increases as the frying progresses (see Figure 3). The concentration of these products in the fumes, at 50 cm from the oil surface, was significant and exceeded 1 ppm (by volume) after several frying cycles. Of the monitored compounds, acrolein was the most and acetaldehyde the least abundant. In the case of 1,3-butadiene, the highest concentration was observed after placing the fifth (and final) batch of fries in the oil.

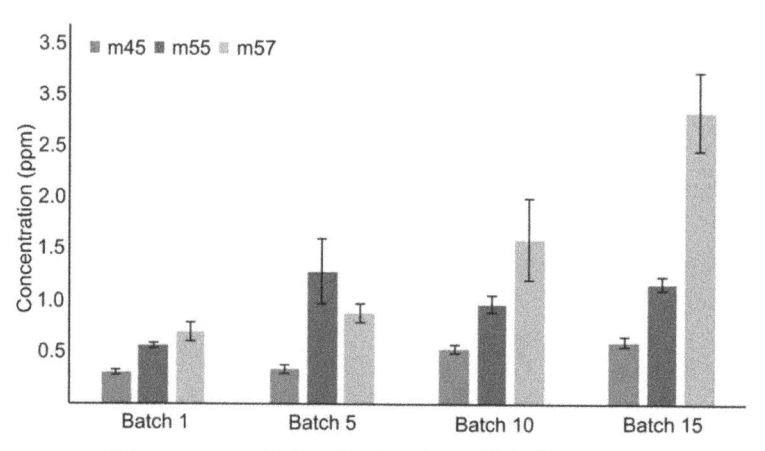

Figure 3. Progression of the emissions of selected VOCs (acetaldehyde, m45; 1,3-butadiene, m55; and acrolein, m57) during frying. Each batch contained ~100 g potato fries, which were immersed into rapeseed oil (2 L) heated to ~160°C. It took ca. 10 min to obtain ready-to-eat fries for each batch before subsequently frying the new batch.

In a situation of prolonged exposure to frying fumes, there may be a risk of poisoning. This may concern mainly people who work professionally in the kitchen or those who prepare food in their households. For example, the US Agency for Toxic Substances and Disease Registry (ATSDR) has determined that the minimum risk level (MRL) for inhalation of acrolein is 0.003 ppm (42). However, this applies to long-term exposure to acrolein vapors. The inlet capillary located at a height of 50 cm represents the worst-case scenario in which the individual preparing the food is leaning over the fryer; from general experience it is known that this occurs only occasionally, mainly when undertaking specific activities, such as stirring, retrieving the fried batch food, or changing the oil.

The fluctuation in the concentrations of these three VOCs was very large, as is evident by the error bars in Figure 3, which represent the standard deviation from second-by-second measurements for the first minute after the batch was placed in the oil. This discrepancy was likely due to the fact that frying was conducted in an open space where uncontrolled mixing of frying fume flux could occur. Additionally, because these volatiles are reactive, secondary reactions may possibly have taken place in the oil fumes. However, this would suggest that for less reactive chemicals, such as saturated aldehydes with a longer carbon chain, the variability should be smaller, which was not the case (see the mentioned publication (23)).

Release Dynamics of Volatiles at Early-Stage Frying

The second study involved deep-frying a potato cube with dimensions of roughly 1.5×1.5×1.5 cm^3 and a mass of 5 g (in most cases; one experiment involved using a potato cube twice as large). The samples were fried in rapeseed oil. Two potato varieties were used in this study. The first, Gala, is a common multi-purpose cultivar popular in Poland. The second, Innovator, is a Dutch cultivar intended for potato fries production. The frying process was conducted at two temperatures, 180°C and 160°C. Temperature measurements indicated that as the potato cube was immersed in oil, the oil temperature dropped dramatically by over 20°C. This is related to the evaporation of water present in the food that acts to temporarily lower the temperature of the oil. The E/N of the instrument was set to 100 Td, and the frying fumes stream was diluted at a ratio of 1:1000 with clean air (total flow of 1000 sccm). The E/N was lower than in the previous experiment in order to reduce the degree of fragmentation of unsaturated compounds, such as alkanals and alkadienals, whereas dilution prevented detector saturation and reduced the influence of water vapor present in the frying fumes.

The early-stage frying experiment involved monitoring the changes in VOC emissions in real time after placing a potato cube in heated rapeseed oil. This section presents part of a study that compared different frying conditions, i.e., different oil temperature, food quantity, and potato cultivar. Both the experimental setup and its working parameters are described in detail elsewhere (17). The focus here, as above, is on the emission of three selected compounds, namely acetaldehyde, 1,3-butadiene, and acrolein. The measurement began 1 min before the potato cube was immersed in the oil and then proceeded for 500 s. At a frying temperature of 180°C, the potato cube was ready to eat after approximately 2.5 min (150 s) of frying, whereas after approximately 7.5 min (450 s) the potato cube was charred and not suitable for consumption. Thus, the whole experiment included the period of oil heating, food immersion, frying, and over-frying.

The act of frying results in a sudden release of VOCs into the gas phase. This can be clearly observed in Figure 4, in which the emission curves of selected VOCs are plotted. The moment of placing food in heated oil (marked as 0 s) initiates a sudden increase in the amount of emitted compounds, whereby the height of the 'bursting peak' depends mainly on the frying temperature and type of food. It is suspected that the key factor causing this effect is the evaporation of water from the surface and bulk of the foodstuff. The higher the temperature, the more intensive the evaporation is, which results in a much narrower and higher bursting peak. The intensity of the bursting peak also depends on the amount of water in the food. Thus, for a potato cultivar dedicated to frying (Innovator with water content 66.5 ± 4.7 %) the bursting peak is much lower than for a general purpose cultivar (Gala; 76.2 ± 2.6 % water). In contrast, the oil-to-food ratio seems to have the least effect on bursting peak formation. The duration of the bursting peak corresponds to the characteristics of water loss during frying, as has been demonstrated in a study by Pedreschi *et al.* in which frying potato slices at 180°C results in the nearly full evaporation of water after approximately 160 s (43).

As soon as the available water is evaporated and a crust is formed, the curve characteristics return to linear growth. The final stage of frying may lead to a plateau, which may be attributed to the reactive nature of these compounds and the occurrence of secondary reactions. At this time, greater signal fluctuations also occur, which may be due to the entry into the over-frying phase and food carbonization processes taking place.

Thus, it appears that the most important role in VOC emissions may be the presence of water in the food. This may be associated with two factors: physical and chemical. In the first case, there is the release of volatiles formed during thermal degradation, which were dissolved in oil. In the second, water, more precisely water vapor, reacts with triacylglycerols, leading to hydrolysis that produces free fatty acids and glycerol (19, 44). However, the key to this phenomenon seems to be the reaction time. Currently, there is a lack of conclusive data on whether hydrolysis plays an important role in the early-stage of frying. Existing studies tend to focus on a period of several hours of frying (45).

Finally, to determine the release kinetics of the selected VOCs from the system, a sound approach is to present the cumulative amount of VOCs released. This approach has been presented several times before, for instance by Mateus *et al.* (46), and provides some guidance on user exposure to chemicals in oil fumes. From the plot in Figure 5 it can be seen that the total emitted amounts of acetaldehyde and acrolein closely follow each other. It can also be noted that the emission of each compound can be described by a quadratic equation, which was also the case with other compounds, such as aldehydes (17).

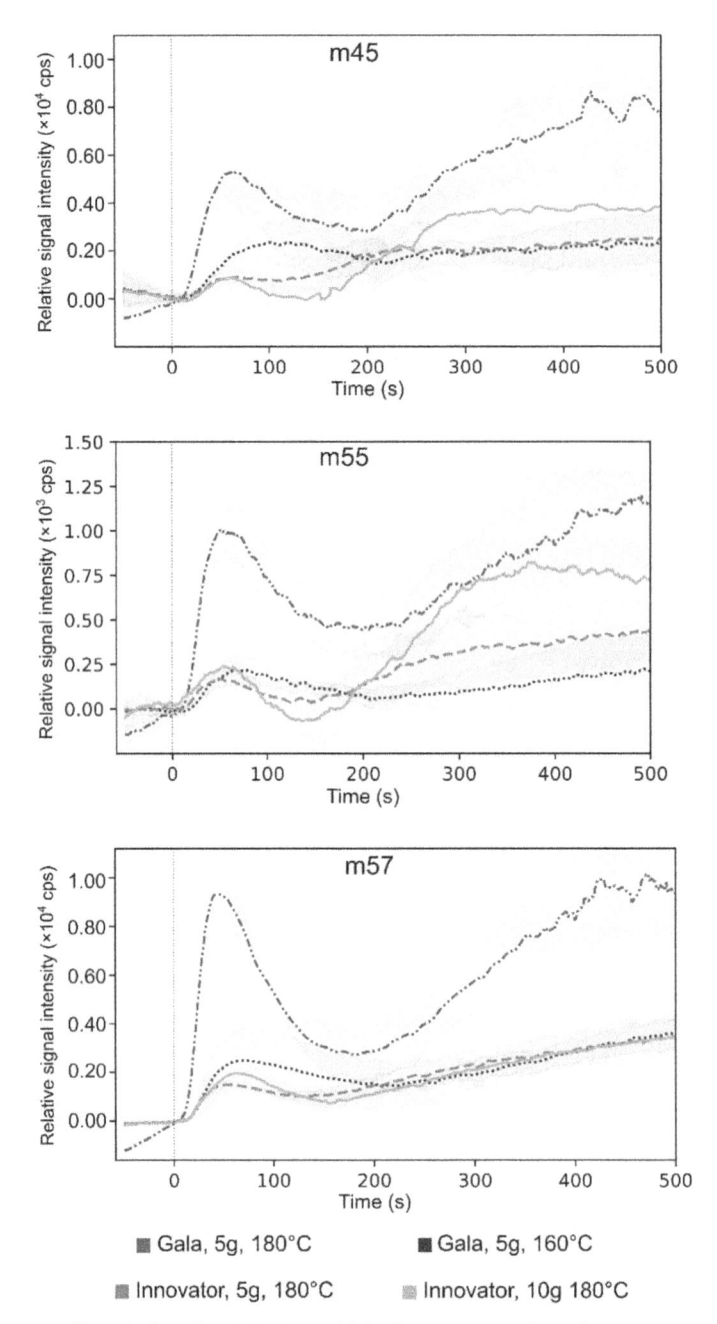

Figure 4. Emission profiles of selected VOCs (acetaldehyde, m45; 1,3-butadiene, m55; and acrolein, m57) generated during potato cube deep frying. Four different scenarios were examined: Gala cultivar potato cube (5 g) immersed into the oil heated to 160°C (blue) or 180°C (violet), or Innovator cultivar potato cube of 5 g (pink) or 10 g (orange) placed into the oil heated to 180°C. Signal intensities are shown as cps relative to time t=0 s (the moment of cube immersion). The signal was smoothed using a moving average (n=10). Shaded areas represent standard deviations (n=3).

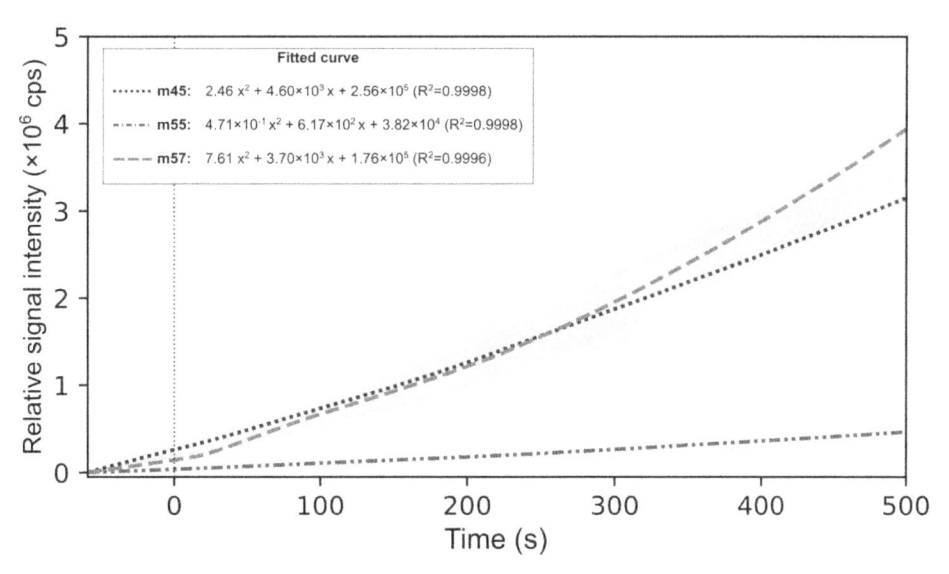

Figure 5. A comparision of the cumulative emission of acetaldehyde (m45), 1,3-butadiene (m55) and acrolein (m57) generated during deep-frying of Innovator cultivar potato cube (5 g) immersed into the rapeseed oil heated to 180°C. The quadratic equation describes the best-fit curve for each VOC. Frying started at time 0 s. Shaded areas represent standard deviation (n=3). Signal intensities are shown as relative cps, where zero is the start of the experiment (1 min before the cube immersion).

Conclusions

Understanding the emission characteristics of VOCs can lead to better insights into the frying process. The processes that take place in the first seconds of frying are particularly interesting. Using DIMS methods it is possible to register early changes in the composition of frying fumes in real time. Using DIMS – in the present case PTR-MS – it is also possible to monitor the concentrations of volatiles that are toxic in nature, including those with carcinogenic character. While short-term exposure to oil vapors is not a threat to health, prolonged exposure to air in which high concentrations of toxic compounds are found can pose a real threat. This applies mainly to people who professionally prepare fried foods. Knowledge of the release kinetics of oil thermal degradation products can be useful in developing guidelines for the proper handling of the frying process. Among others, the frying temperature and the amount of water delivered with the food can play an important role. In the future, it is important to determine the changes in water content in frying fumes and in fried foodstuffs. This would make it possible to directly determine the influence of water vapor on the formation and emission of volatile SOPs.

Further research should be directed towards the analysis of other food-oil systems, which would lead to an understanding of the replationships between the composition of food, oil type and the emission profile of harmful volatiles.

Acknowledgments

This research was funded by the National Science Centre, Poland, grant number 2018/31/N/NZ9/02404.

References

1. Ibáñez, C.; Simó, C.; García-Cañas, V.; Acunha, T.; Cifuentes, A. The role of direct high-resolution mass spectrometry in foodomics. *Anal Bioanal Chem* **2015**, *407*, 6275–6287.
2. Biasioli, F.; Gasperi, F.; Yeretzian, C.; Märk, T. D. PTR-MS monitoring of VOCs and BVOCs in food science and technology. *TrAC - Trends Anal Chem* **2011**, *30*, 968–977.
3. Biasioli, F.; Yeretzian, C.; Märk, T. D.; Dewulf, J.; Van Langenhove, H. Direct-injection mass spectrometry adds the time dimension to (B)VOC analysis. *TrAC - Trends Anal Chem* **2011**, *30*, 1003–1017.
4. Wojnowski, W.; Majchrzak, T.; Szweda, P.; Dymerski, T.; Gębicki, J.; Namieśnik, J. Rapid Evaluation of Poultry Meat Shelf Life Using PTR-MS. *Food Anal Methods* **2018**, 1–8.
5. Alothman, M.; Lusk, K. A.; Silcock, P.; Bremer, P. J. Comparing PTR-MS profile of milk inoculated with pure or mixed cultures of spoilage bacteria. *Food Microbiol* **2017**, *64*, 155–163.
6. Kuuliala, L.; Sader, M.; Solimeo, A.; Pérez-Fernández, R.; Vanderroost, M.; De Baets, B.; De Meulenaer, B.; Ragaert, P.; Devlieghere, F. Spoilage evaluation of raw Atlantic salmon (Salmo salar) stored under modified atmospheres by multivariate statistics and augmented ordinal regression. *Int J Food Microbiol* **2019**, *303*, 46–57.
7. Ozcan-Sinir, G.; Copur, O. U.; Barringer, S. A. Botanical and geographical origin of Turkish honeys by selected-ion flow-tube mass spectrometry and chemometrics. *J Sci Food Agric* **2020**, *100*, 2198–2207.
8. Araghipour, N.; Colineau, J.; Koot, A.; Akkermans, W.; Rojas, J. M. M.; Beauchamp, J.; Wisthaler, A.; Märk, T. D.; Downey, G.; Guillou, C.; Mannina, L.; van Ruth, S. Geographical origin classification of olive oils by PTR-MS. *Food Chem* **2008**, *108*, 374–383.
9. Schueuermann, C.; Bremer, P.; Silcock, P. PTR-MS volatile profiling of Pinot Noir wines for the investigation of differences based on vineyard site. *J Mass Spectrom* **2017**, *52*, 625–631.
10. Acierno, V.; Liu, N.; Alewijn, M.; Stieger, M.; van Ruth, S. M. Which cocoa bean traits persist when eating chocolate? Real-time nosespace analysis by PTR-QiToF-MS. *Talanta* **2019**, *195*, 676–682.
11. Tyapkova, O.; Siefarth, C.; Schweiggert-Weisz, U.; Beauchamp, J.; Buettner, A.; Bader-Mittermaier, S. Flavor release from sugar-containing and sugar-free confectionary egg albumen foams. *LWT - Food Sci Technol* **2016**, *69*, 538–545.
12. Charles, M.; Romano, A.; Yener, S.; Barnabà, M.; Navarini, L.; Märk, T. D.; Biasoli, F.; Gasperi, F. Understanding flavour perception of espresso coffee by the combination of a dynamic sensory method and in-vivo nosespace analysis. *Food Res Int* **2015**, *69*, 9–20.
13. Dimitri, G.; van Ruth, S. M.; Sacchetti, G.; Piva, A.; Alewijn, M.; Arfelli, G. PTR-MS monitoring of volatiles fingerprint evolution during grape must cooking. *LWT - Food Sci Technol* **2013**, *51*, 356–360.
14. Klein, F.; Farren, N. J.; Bozzetti, C.; Daellenbach, K. R.; Kilic, D.; Kumar, N. K.; Pieber, S. M.; Slowik, J. G.; Tuthill, R. N.; Hamilton, J. F.; Baltensperger, U.; Prévôt, A. S. H.; El Haddad, I. Indoor terpene emissions from cooking with herbs and pepper and their secondary organic aerosol production potential. *Sci Rep* **2016**, *6*, 1–7.
15. Gloess, A. N.; Vietri, A.; Wieland, F.; Smrke, S.; Schönbächler, B.; López, J. A. S.; Petrozzi, S.; Bongers, S.; Koziorowski, T.; Yeretzian, C. Evidence of different flavour formation dynamics by

roasting coffee from different origins: On-line analysis with PTR-ToF-MS. *Int J Mass Spectrom* **2014**, *365–366*, 324–337.

16. Pico, J.; Khomenko, I.; Capozzi, V.; Navarini, L.; Biasioli, F. Real-time monitoring of volatile compounds losses in the oven during baking and toasting of gluten-free bread doughs: A PTR-MS evidence. *Foods* **2020**, *9*, 1498.
17. Majchrzak, T.; Wasik, A. Release Kinetics Studies of Early-Stage Volatile Secondary Oxidation Products of Rapeseed Oil Emitted during the Deep-Frying Process. *Molecules* **2021**, *26*, 1006.
18. Choe, E.; Min, D. B. Mechanisms and factors for edible oil oxidation. *Compr Rev Food Sci Food Saf* **2006**, *5*, 169–186.
19. Choe, E.; Min, D. B. Chemistry of deep-fat frying oils. *J Food Sci* **2007**, *72*, 1–10.
20. Bravo, J.; Sanjuán, N.; Ruales, J.; Mulet, A. Modeling the dehydration of apple slices by deep fat frying. *Dry Technol* **2009**, *27*, 782–786.
21. Lojzova, L.; Riddellova, K.; Hajslova, J.; Zrostlikova, J.; Schurek, J.; Cajka, T. Alternative GC-MS approaches in the analysis of substituted pyrazines and other volatile aromatic compounds formed during Maillard reaction in potato chips. *Anal Chim Acta* **2009**, *641*, 101–109.
22. Pedreschi, F.; Enrione, J. Frying of Foods. In *Conventional and Advanced Food Processing Technologies*; Wiley Blackwell: 2014; pp 197–220.
23. Majchrzak, T.; Wojnowski, W.; Głowacz-Różyńska, A.; Wasik, A. On-line assessment of oil quality during deep frying using an electronic nose and proton transfer reaction mass spectrometry. *Food Control* **2021**, *121*, 107659.
24. Frankel, E. N. Volatile lipid oxidation products. *Prog Lipid Res* **1983**, *22*, 1–33.
25. Ahmed, M.; Pickova, J.; Ahmad, T.; Liaquat, M.; Farid, A.; Jahangir, M. Oxidation of Lipids in Foods. *Sarhad J Agric* **2016**, *32*, 230–238.
26. Shields, P. G.; Xu, G. X.; Blot, W. J.; Fraumeni, J. F.; Trivers, G. E.; Pellizzari, E. D.; Qu, Y. H.; Gao, Y. T.; Harris, C. C. Mutagens from heated chinese and U.S. Cooking oils. *J Natl Cancer Inst* **1995**, *87*, 836–841.
27. Tarawneh, I. N.; Najjar, A. A.; Bani Issa, R. S.; Salameh, F. F.; Abu Shmeis, R. M. Determination of Polycyclic Aromatic Hydrocarbons and α,β-Unsaturated Aldehydes in Frying Oils in Jordan. *Polycycl Aromat Compd* **2020**.
28. Huang, Y.; Ho, S. S. H.; Ho, K. F.; Lee, S. C.; Yu, J. Z.; Louie, P. K. Characteristics and health impacts of VOCs and carbonyls associated with residential cooking activities in Hong Kong. *J Hazard Mater* **2011**, *186*, 344–351.
29. Wang, L.; Csallany, A. S.; Kerr, B. J.; Shurson, G. C.; Chen, C. Kinetics of Forming Aldehydes in Frying Oils and Their Distribution in French Fries Revealed by LC-MS-Based Chemometrics. *J Agric Food Chem* **2016**, *64*, 3881–3889.
30. Fullana, A.; Carbonell-Barrachina, Á. A.; Sidhu, S. Volatile aldehyde emissions from heated cooking oils. *J Sci Food Agric* **2004**, *84*, 2015–2021.
31. Bein, K.; Leikauf, G. D. Acrolein - a pulmonary hazard. *Mol Nutr Food Res* **2011**, *55*, 1342–1360.
32. Henning, R. J.; Johnson, G. T.; Coyle, J. P.; Harbison, R. D. Acrolein Can Cause Cardiovascular Disease: A Review. *Cardiovasc Toxicol* **2017**, *17*, 227–236.
33. Leikauf, G. D. Hazardous air pollutants and asthma. *Environ Health Perspect* **2002**, *110* (Suppl), 505–526.

34. Stevens, J. F.; Maier, C. S. Acrolein: Sources, metabolism, and biomolecular interactions relevant to human health and disease. *Mol Nutr Food Res* **2008**, *52*, 7–25.
35. Salaspuro, M. Acetaldehyde and gastric cancer. *J Dig Dis* **2011**, *12*, 51–59.
36. Pellizzari, E. D.; Michael, L. C.; Thomas, K. W.; Shields, P. G.; Harris, C. Identification of 1,3-butadiene, benzene, and other volatile organics from wok oil emissions. *J Expo Anal Environ Epidemiol* **1995**, *5*, 77–87.
37. Tong, R.; Zhang, B.; Yang, X.; Cao, L. Health risk assessment of chefs intake of cooking fumes: Focusing on Sichuan cuisine in China. *Hum Ecol Risk Assess* **2021**, *27*, 162–190.
38. Nakamura, J.; Carro, S.; Gold, A.; Zhang, Z. An unexpected butadiene diolepoxide-mediated genotoxicity implies alternative mechanism for 1,3-butadiene carcinogenicity. *Chemosphere* **2021**, 266–129149.
39. Ellis A. M., Mayhew C. A. *Proton Transfer Reaction Mass Spectrometry*; Wiley: 2014.
40. Blake, R. S.; Monks, P. S.; Ellis, A. M. Proton-transfer reaction mass spectrometry. *Chem Rev* **2009**, *109*, 861–896.
41. Jordan, A.; Haidacher, S.; Hanel, G.; Hartungen, E.; Märk, L.; Seehauser, H.; Schottkowsky, R.; Sulzer, P.; Märk, T. D. A high resolution and high sensitivity proton-transfer-reaction time-of-flight mass spectrometer (PTR-TOF-MS). *Int J Mass Spectrom* **2009**, *286*, 122–128.
42. Agency for Toxic Substances and Disease Registry (ATSDR). *Toxicological Profile for Acrolein*; Atlanta, GA, 2007.
43. Pedreschi, F.; Hernández, P.; Figueroa, C.; Moyano, P. Modeling water loss during frying of potato slices. *Int J Food Prop* **2005**, *8*, 289–299.
44. Bordin, K.; Kunitake, M. T.; Aracava, K. K.; Trindade, C. S. F. Changes in food caused by deep fat frying - A review. *Arch Latinoam Nutr* **2013**, *63*, 5–13.
45. Dana, D.; Blumenthal, M. M.; Saguy, I. S. The protective role of water injection on oil quality in deep fat frying conditions. *Eur Food Res Technol* **2003**, *217*, 104–109.
46. Mateus, M. L.; Lindinger, C.; Gumy, J. C.; Liardon, R. Release kinetics of volatile organic compounds from roasted and ground coffee: Online measurements by PTR-MS and mathematical modeling. *J Agric Food Chem* **2007**, *55*, 10117–10128.

Editor's Biography

Jonathan D. Beauchamp

Jonathan Beauchamp holds a physics degree from University College London, UK (2002) and a doctoral degree in environmental physics from the University of Innsbruck, Austria (2005). Since 2008, he has held a position as research associate at the Fraunhofer Institute for Process Engineering and Packaging IVV in Freising, Germany, where he currently manages the Dynamic Emissions Analytics and Diagnostics group and is Deputy Head of the Department of Sensory Analytics and Technologies. Jonathan's research encompasses the characterization of emissions of volatile organic compounds (VOCs) in the fields of food and flavor, non-consumer goods, and the human volatilome, the latter primarily directed towards the detection of volatile disease biomarkers in exhaled breath. His further scientific interests and activities relate to the study of odor-active compounds and their role in human olfaction. Jonathan's technical expertise center on the use of real-time mass spectrometry in the form of proton transfer reaction-mass spectrometry (PTR-MS) for analyzing dynamic volatile emissions, but extend to other techniques, including (comprehensive) gas chromatography mass spectrometry with olfactometry (GC-MS/O and GC×GC-MS) and ion mobility spectrometry (GC-IMS).

Jonathan has over 70 papers and book chapters to his name, and has co-edited two previous volumes within this ACS Symposium Series; *The Chemical Sensory Informatics of Food: Measurement Analysis and Integration* (2015) and *Sex, Smoke, and Spirits: The Role of Chemistry* (2019). He has co-organized and chaired several ACS symposia and has been on the organizing committee of other international conferences, notably the PTR-MS Conference series and the International Association of Breath Research (IABR) Breath Summits. Jonathan has been an full member of ACS and has actively engaged in the Division of Agricultural and Food Chemistry (AGFD) since 2013, where he currently serves as the Flavor sub-division chair. He is affiliated with several scientific journals, including *Journal of Breath Research* (Associate Editor), *Food Packaging and Shelf Life* (Editorial Board Member), and *Heliyon Environment* (Advisory Board Member).

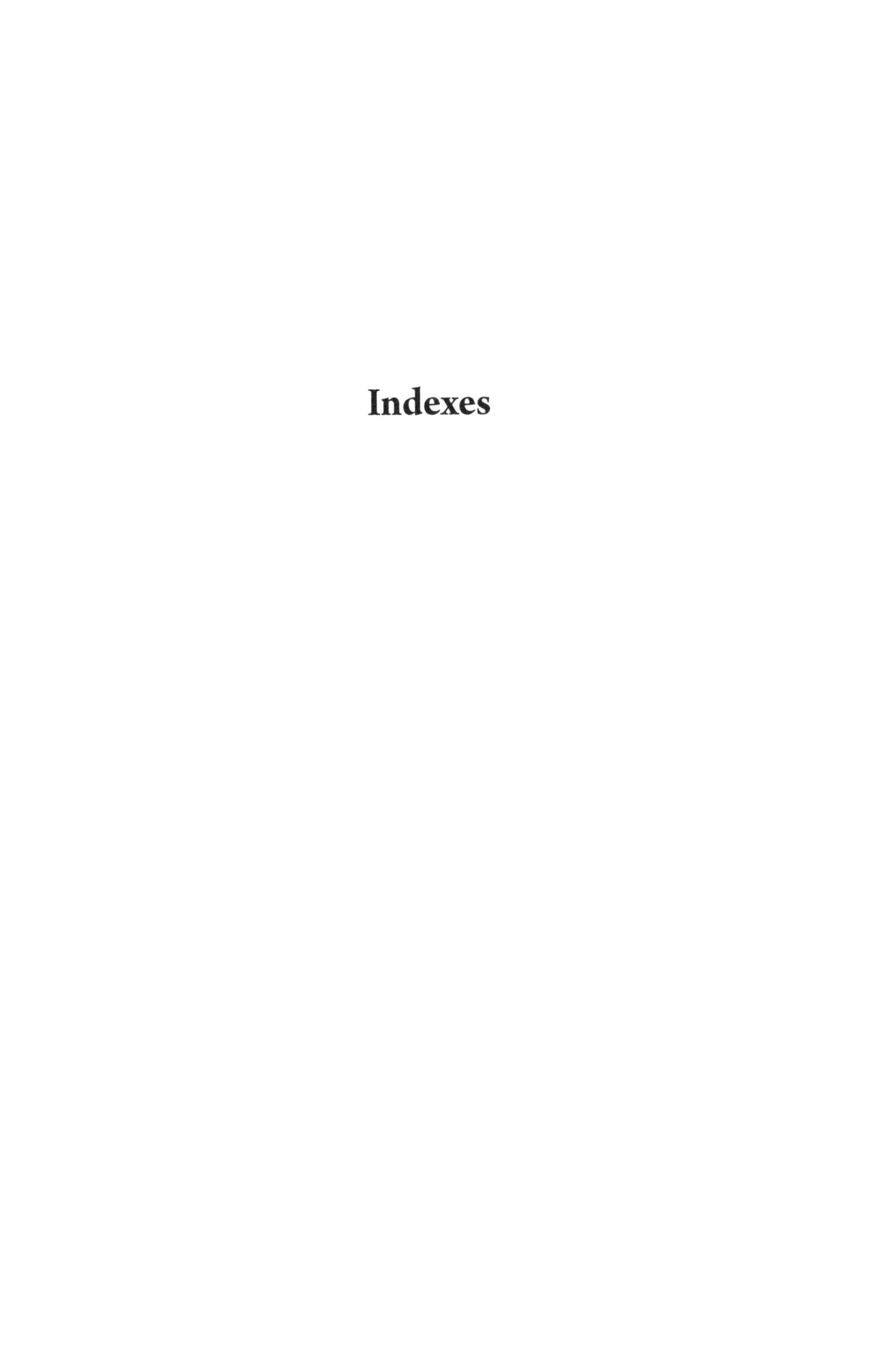

Indexes

Author Index

Barringer, S., 99
Beauchamp, J., xi, 1, 33
Biasioli, F., 123
Canon, F., 67
Capozzi, V., 123
De Baets, B., 107
Devlieghere, F., 107
Fisk, I., 87
Ford, C., 87
Fragasso, M., 123
Jain, N., 107
Khomenko, I., 123
Kuuliala, L., 107
Langford, V., 1, 51
Majchrzak, T., 139
Muñoz-González, C., 67
Padayachee, D., 51
Pozo-Bayón, M., 67
Silcock, P., 123
Taylor, A., 1, 17, 77
Wichrowska, W., 139
Yabuki, M., 77
Yang, N., 87

Subject Index

A

Acrolein, acetaldehyde and 1,3-butadiene, monitoring
   conclusions, 147
   direct injection mass spectrometry, 139
   frying, toxic volatile products, 140
      fumes, PTR-MS mass spectral fingerprints, 142f
   monitoring acetaldehyde, 1,3-butadiene, 142
      PTR-MS analysis, two sample delivery methods, 143f
   volatile emissions during prolonged frying, 143
      selected VOCs, progression of the emissions, 144f
   volatiles at early-stage frying, release dynamics, 144
      cumulative emission of acetaldehyde, comparision, 147f
      selected VOCs, emission profiles, 146f
APCI-MS/MS
   conclusions, 96
   experimental methods, 89
      data analysis, 91
      different physicochemical properties, nine selected aroma compounds, 90t
   introduction, 87
      triple quadrupole mass spectrometer, schematic, 88f
   results and discussion, 92
      anisole and 2,5-dimethyl pyrazine, imax values, 94f
      anisole and 2,5-dimethyl pyrazine, precursor fragmentation, 93f
      2,5-dimethyl pyrazine, mean in-nose aroma intensity, 95f
      nine volatile compounds, calculated signal-to-noise (S/N) ratios, 92f
Aroma persistence using real-time mass spectrometry, molecular basis
   aroma persistence, phenomenon, 68
      publications exploring the phenomenon of aroma persistence, overview, 69t
   conclusions, 73
   factors underpinning aroma persistence, 70
      aroma persistence, role of food matrix composition, 72
      physiological mechanisms, hypothesis, 71f
   introduction, 67

D

Different food matrices using PTR-MS, real-time monitoring of flavoring starter cultures, 123
   conclusions, 133
      tailored green analytical techniques, 134
   experimental methods, 125
   introduction, 124
      main DIMS analytical approaches, principal features characterizing, 124t
   results and discussion, 126
      chosen m/z associated with dough fermentation, variability in the time evolution, 130f
      detection of VOCs, PTR-TOF-MS applied, 127t
      different combinations of yeast and flour, variability of mass peak, 131f
      dough, average mass spectra, 129f
      PCA, score plot, 132f
      PCA of VOCs, score plot, 128f
      S. cerevisiae bakery starter, leavening process, 133t
Direct APCI-MS, flavor applications
   conclusion, 26
      APCI applications, 26
   plant breeding studies, self-service APCI-MS system, 25
      releasing tomato aromas for APCI-MS analysis, maceration device, 26f
   understanding dynamic flavor systems using direct APCI-MS, 22
      first stage of the Maillard reaction, chemical pathways involved, 24f
      headspace dilution, plot of headspace ion intensity against time, 23f
   in vivo aroma release analyses, learnings, 17
      aroma release, publications, 18t

aroma release variation in human panelists, dealing, 21
direct APCI-MS, dealing with data, 22
GC-APCI-MS selected ion traces, 20f
log lipid effect for regular and low fat milk samples, frequency distribution, 22f
monitoring volatile release from tea during brewing, shematic of a system, 20f

Dynamic flavor analysis with PTR-MS, pushing the boundaries
conclusions, 42
food/flavor analysis, heralding a new era, 35
beginnings, 33
outside-the-box PTR-MS highlights, 38
PTR-MS repertoire, expanding, 41
PTR-TOF-MS, 39

M

Methods of choice for exploring flavor release, APCI-MS, PTR-MS and SIFT-MS
APCI-MS, 8
conclusions, 12
PTR-MS, 9
real-time mass spectrometry, rationale, 1
real-time mass spectrometry methods, overview, 3
APCI-MS, PTR-MS and SIFT-MS instrumentation, generic configurations, 7f
APCI-MS, PTR-MS and SIFT-MS real-time mass spectrometry systems, key features, 4t
common classes of volatile (aroma) compounds, proton affinity values, 5f
three most common reagent ions, reaction mechanisms, 6t
SIFT-MS, 11

S

Shredded carrot using SIFT-MS, identifying potential volatile spoilage indicators
conclusions, 119
experimental methods, 108
analysis of shredded carrot, VOCs and SIFT-MS parameters, 110t
introduction, 107
results and discussion, 113
cross-validation models, perplexities, 115t
extracted profiles, distribution of top eight VOCs, 117f
headspace gases, evolution, 114f
profiles in samples, distribution, 116f
selected VOCs, concentrations, 118f

SIFTing through flavor
chemometrics and SIFT-MS, 52
conclusions, 62
introduction, 51
SIFT-MS, industrial applications, 53
function of temperature, for whole milk powder, VOC and CO concentration traces, 60f
paper classification, most discriminating compounds, 61f
Parmigiano Reggiano cheese, classification, 57f
Parmigiano Reggiano cheeses, manufacturer and sample information, 56t
real-time food processing, 58
strawberry-flavored standards and unknowns, classification, 55f

Simulated peri-receptor conditions, using APCI-MS to study the dynamics of odor binding
conclusions, 83
explore odorant interactions with OBP, developing in vitro models, 79
IBT binding displaces IBMP, 83f
OBP-bound isobutylthiazole, release, 82f
off-rate experiment for n-butyl acetate, APCI-MS traces, 79f
peri-receptor region, conditions, 81t
presence of isobutylthiazole, APCI-MS trace monitoring, 81f
in vitro OBP system, schematic, 80f
odor binding protein, role, 77
odors after intranasal administration, recognition, 78t

U

Using SIFT-MS to improve food quality, from mold worms to fake honey
adulteration, 103
honey adulterated with corn syrup, principal component analysis showing differentiation of pure honey, 104f
classification, 104

varieties of garlic, principal component analysis showing differentiation, 104*f*
conclusions, 104
source of food items, adulteration and authenticating, 104
off-odor in fruit juice, determination of the cause, 99
A. acidoterrestris, growth, 101*f*
A. acidoterrestris, growth and guaiacol production, 101*f*
vanillin concentration and amount of guaiacol produced, relationship between, 100*f*
vanillin into guaiacol, proposed conversion, 100*f*
Swiss cheese, understanding consumer acceptance, 101
differentiation of Swiss cheeses, principal component analysis, 102*f*